Studies in Computational Intelligence 449

Editor-in-Chief

Prof. Janusz Kacprzyk
Systems Research Institute
Polish Academy of Sciences
ul. Newelska 6
01-447 Warsaw
Poland
E-mail: kacprzyk@ibspan.waw.pl

For further volumes:
http://www.springer.com/series/7092

Muhammet Ünal, Ayça Ak, Vedat Topuz,
and Hasan Erdal

Optimization of PID Controllers Using Ant Colony and Genetic Algorithms

 Springer

Authors

Muhammet Ünal
Technical Education Faculty, D401
Marmara University
Goztepe Campus
Istanbul
Turkey

Vedat Topuz
Vocational School of Technical Sciences
Marmara University
Goztepe Campus
Istanbul
Turkey

Ayça Ak
Vocational School of Technical Sciences
Marmara University
Goztepe Campus
Istanbul
Turkey

Hasan Erdal
Technology Faculty
Marmara University
Goztepe Campus
Istanbul
Turkey

ISSN 1860-949X
ISBN 978-3-642-43477-8
DOI 10.1007/978-3-642-32900-5
Springer Heidelberg New York Dordrecht London

e-ISSN 1860-9503
ISBN 978-3-642-32900-5 (eBook)

Printed on acid-free paper

Springer is part of Springer Science+Business Media (www.springer.com)

Foreword

This book describes a real time control algorithm using Genetic Algorithm (GA) and Ant Colony Optimization (ACO) algorithm for optimizing PID controller parameters. Proposed method was tested on GUNT RT 532 Pressure Process Control System. The dynamic model of the process to be controlled was obtained using Artificial Neural Network (ANN). Using the chosen model, the parameters of PID controller were optimized with ACO, GA and Ziegler-Nichols (ZN) techniques. The performances of these three techniques were compared with each other using the criteria of overshoot, rise time, settling time and root mean square (RMS) error of the trajectory. It was observed that the performances of GA and ACO are better than that of ZN technique.

This book was prepared based on the Muhammet UNAL's master thesis entitled **Optimization of PID Controller Using Ant Colony / Genetic Algorithms and Control of The GUNT RT 532 Pressure Process** at Marmara University Institute for Graduate Studies in Pure and Applied Sciences.

Contents

List of Symbols

W	: Artificial Neural Network's Weight Matrix
f	: Artificial Neural Network's Activation Function
p	: Artificial Neural Network's Input
a	: Artificial Neural Network's Output
d	: Artificial Neural Network's Desired Response
λ	: Gain Factor of the Continuous Activation Function
b	: Artificial Neural Network's Bias Value
X	: Input Vector of Pattern
T	: Corresponding Output Target
Y	: Output From the Neuron
δ	: Error Value from Output to Input Layer
β	: Weight Adjustment Factor Normalized Between 0 and 1
E	: Error Matrix
O	: Artificial Neural Network's Obtain Output Values
$\hat{y}$	: ANN''s Predicted Value
y	: Process output
u	: Process Input
e	: Error value
ε	: Approximation Error
$F(.)$	: Multi-dimensional System Definition
$N(.)$	: Nonlinear Definition of the ANN

D : Time Delay

t : Time

Σ : Search Space (phenotype) of Genetic Algorithm

G : Coding Space (genotype) of Genetic Algorithm

ρ : Coding Function ($\rho:\Sigma\rightarrow G$) of Genetic Algorithm

f : Fitness Function ($f:\Sigma\rightarrow R^+$) of Genetic Algorithm

μ : Population Size of Genetic Algorithm

I : Initialization Function ($I:\mu\rightarrow P(G)$) of Genetic Algorithm

$P(G)$: Population of Genetic Algorithm

S : Selection Type of Genetic Algorithm

Gap : Percentage of the Population that Conveyed from Old Generation to New Generation of Genetic Algorithm

Ω : Genetic Operator of Genetic Algorithm

ω_c : Crossover Operator of Genetic Algorithm

ω_m : Mutation Operator of Genetic Algorithm

R : Replacement Policy of Genetic Algorithm

τ : Termination Criterion of Genetic Algorithm

Pm : Mutation Probability

Pc : Crossing Probability

$O_{(S)}$: Grade of Scheme

L : Length of the Bit String

$\delta(S)$: Length of Scheme

$\xi_{(S,t)}$: Number of Series Matching with Scheme S in t Time

$\Phi_{(S,t)}$: Fitness of S Scheme in t Time

$\Phi_{(t)}$: Approximate Value of Fitness

O_o : Initial Number of Population Scheme

O_c : Number of Population Schemes which are Converging

T_C : Time Until Convergence

q : Size of the Tournament

T_G : Number of Generations in a Moment

M : Maximum Number of Generations

ϕ : Fitness Value

σ : Standard Deviation of the Population

$\tau_{ij}(t)$: At Time t Trace Amount of Pheromones at (i,j) the Corners.

η_{ij} : Visibility value between (i, j) corners.

α : Relative Importance of the Pheromone Trace in the Problem

β : Importance that Given to Visibility Value

N_i : Set of the Node Points that hasn't Chosen Yet

$\Delta\tau_{ij}$: Amount of Pheromone Trace of the Corner Due to the Election of the (i, j) Corner During a Tour of the Ant.

m : Total Ant Number

$\Delta\tau_{ij}^{k}$: Amount of Pheromone Trace Left by k. Ant at (i, j) Corner

Q : Constant

L_k : Tour Length of k. Ant

P_0 : Initial Pressure Value

P_s : Desired Pressure Value

$\varphi_{0\text{-}3}$: Trajectory Coefficients

t_o : Initial Time

t_S : Setting Time

u : Control Signal

List of Abbreviations

PID	:	Proportional Integral Derivative
SC	:	Soft Computing
FL	:	Fuzzy Logic
ANN	:	Artificial Neural Network
MLP	:	Multilayer Perceptron
NARX	:	Nonlinear Autoregressive Network with Exogenous Inputs
TDL	:	Time Delay Layer
SL	:	Supervised Learning
UL	:	Unsupervised Learning
GA	:	Genetic Algorithm
SGA	:	Simple Genetic Algorithm
μGA	:	Micro Genetic Algorithm
ssGA	:	Steady State Genetic Algorithm
HGA	:	Hierarchic Genetic Algorithm
mGA	:	Messy Genetic Algorithm
SSRS	:	Stochastic Sampling with Replacement Selection
DNA	:	Deoxyribonucleic Acid
SUS	:	Stochastic Universal Sampling
ACO	:	Ant Colony Optimization
DAQ	:	Data Acquisition
ADC	:	Analog Digital Converter
DAC	:	Digital Analog Converter
MSE	:	Mean Square Error
RMS	:	Root Mean Square
MSEREG	:	Mean Squared Error with Regularization
SSE	:	Sum Square Error
SISO	:	Single Input and Single Output
ZN	:	Ziegler-Nichols
GMVC	:	Generalization Minimum Variable Control

List of Figures

List of Tables

Introduction

This book is about trajectory tracking control of a pressure control system by PID controller with modeling Artificial Neural Networks (ANN) and finding the coefficients of PID controller with Genetic Algorithm (GA), Ant Colony Algorithm (ACO) and the Ziegler-Nichols (ZN). This book also presents performance comparison of these methods. This book was prepared based on the master thesis entitled Optimization of PID Controller Using Ant Colony / Genetic Algorithms and Control of The GUNT RT 532 Pressure Process at Marmara University Institute for Graduate Studies in Pure and Applied Sciences [1].

There are many researches and studies done on control applications dealing with pressure systems and ANN, GA, Genetic-PID controller, the ACO and the Ant-PID controllers used in control phase in the literature. Although there are experimental methods for the control of a system, the mathematical model of the system is needed to perform especially simulation-based studies. Obtaining mathematical models of linear systems whose parameters belonging to dynamic behavior are known is easy. But, obtaining mathematical models of nonlinear systems is quite difficult and time consuming too. In such cases, methods based input-output behavior of the system is used. Modeling with ANN is the most preferred one of these methods.

H. Mok et al. have controlled a nonlinear pressure control system with the time delay by digital PI controller. System model is extracted as a first order transfer function with time delay and controllers were optimized using ZN method according to open-loop frequency response [2].

Because of finding the coefficients of the PID controller properly will be difficult due to the modeling error and problems resulting from changing the system over time, R.Namba et al. have performed controlling the air pressure unit by determining the parameters of the PID controller according to the reinforced robust stability theory [3].

C. Guo, has practiced control of heating, ventilation and air conditioning (HVAC) system by ANNs supported with PI controller. They have used online-trained ANN to eliminate the noise in the system. In this system, researchers can send controlled signal to the system even error during the control is zero so they can hold the system at the reference level owing to PI controller [4].

In the study performed by W. Jian et al, control of air pressure unit for the air conditioning system is realized. During the operation of the system with PID controller, the system output properly continues within the tolerance. However, the system with time-varying parameters has begun to show nonlinear behavior. In this case the PID controller remained weak to control the system and researchers have developed non- linear controllers to address this weakness. For this purpose,

M. Ünal et al.: Optimization of PID Controllers Using AC and GA, SCI 449, pp. 1–3.
springerlink.com   © Springer-Verlag Berlin Heidelberg 2013

a model of the system with the second-order and time delay was revealed. PID and fuzzy controller are optimized by researchers and neuro-fuzzy controller is optimized by mean least squares, then system is controlled by three controllers. It's shown that results obtained from neuro-fuzzy controller are more successful [5].

M. Unal et al, designed PID controller with GA to track ISAC (Intelligent Soft Arm Control) robot arm whose transfer function is known a reference trajectory in simulation environment [6].

M. Salami et al realized detecting the gain value of the PID controller and optimization the PID coefficients that provide diminution of the total error of system over transfer function. When necessary parameters for the GA are encoded, binary encoding method is used [7].

Jones et al chose controlling the temperature of laboratory as a system that will be controlled and firstly model of the system was created. In the second stage, they found the coefficients of the PID controller for controlling of the system on offline process according to the principle of minimizing the total error with GA. Finally, they compared performance of PID controller coefficients founded by GA with Ästrom -Relay Auto Tuning method and they demonstrated that coefficients founded by GA have better results [8].

Ö. Gundogdu made a study about optimal tuning of the PID controller used in the control of a mass with an inertia driven gear box. In this study, the optimal controller parameters were found with GA by achieving the RMS (Root Mean Square) mistake between the steady state values of the system output and the reference value. In the study, ZN method and GA results were compared. Better results were obtained using GA [9].

Dandil et al designed a model reference PID controller with GA for a variety of linear and nonlinear systems. They found most appropriate PID parameters that provide desired control criteria for many linear and nonlinear systems [10].

H.Alli et al optimized PID control parameters using GA according to different performance indexes for two dynamic systems. Also they showed that most appropriate performance indexes are Integral of the Time Absolute Error (ITAE) [11].

Mitsukura Y. et al designed the optimization of PID controller that optimized the unknown or time varying system parameters. In the study, PID controller and optimization system based on rules of the Generalized Minimum Variance Control were realized. Least squares optimization method was used in calculating the adaptation value in optimization. At the same time, they showed that searching time decreases by reducing the GA search space with user knowledge [12].

H. Duan et al has ran a flight Simulator whose PID controller coefficients were calculated by ACO and optimization of the PID controller was realized by taking the difference between the reference and the output [13].

In study of T. Ergüzel et al, action values of the fuzzy control with ACO and coefficient values of the PID controller were found. Minimization of the RMS error was tried while the controller was optimizing [14].

H. Varol et al designed a cost function and calculated the total absolute error and total square error through the transfer function given during the optimization of the PID controller with ACO. Results with less settlement time, very little or no overshoot were achieved by PID controller optimized by ACO. The performance was measured by comparing the obtained results with Internal Model Control (IMC) and Iterative Feedback Tuning (IFT) [15].

Unlike the above-mentioned studies, this book describes the following issues:

Optimization of the PID controller's coefficients for process (pressure tank) control was found by GA and ACO through ANN model. In real-time study, performance of the PID created controller was observed.

The method that ensured finding the minimum RMS value of the error between the reference and PID controller output during the finding of the PID controller parameters by GA and ACO was analyzed. It tried the working of the PID controller designed in this way according to the desired performance criterion on different input levels.

System has shown a nonlinear behavior at situations like: compressibility of the air, changing of the pressure with air temperature, varied environment temperatures and disruptive effects in pressure control system. For these reasons, ANN model was used instead of mathematical model for model of the pressure system. The behavior of the system was recorded by giving inputs with different values to the pressure system and these input-output data was applied to the ANN as training data. Then the dynamic model of the system was obtained.

The optimal values of coefficients of the PID controller were found with GA and ACO through obtained dynamic neural network model. Minimization of the total absolute error between the reference and the output of the system was selected as objective / optimality criterion. System was controlled on real time with the founded PID coefficients. Different steps and trajectory input values were given to test the real-time behavior of the system. The resulting system behaviors were compared with each other and the PID controller setting by ZN method.

This book is arranged as follows: ANN is introduced in the next section. Section 2 and section 3 contains Genetic Algorithm and Ant Colony Optimization respectively. An application for process system control is presented in section 4. In this section, GUNT RT 532 Pressure Process Control System, PID Controller Design with GA and PID Controller Design with ACO are explained. Section 5 concludes the book. ACO, Cost Function, and codes for operation of these on Matlab are given on annex.

Chapter 1
Artificial Neural Networks

An Artificial Neural Network is an information processing system that has certain performance characteristics in common with biological neural networks. Artificial neural networks have been developed as generalizations of mathematical models of human cognition or neural biology [16]. The basic processing elements of neural networks are called neurons. Neuron performs as summing and nonlinear mapping junctions. In some cases they can be considered as threshold units that fire when their total input exceeds certain bias levels. Neurons usually operate in parallel and are configured in regular architectures. The neurons are generally arranged in parallel to form layers. Strength of the each connection is expressed by a numerical value called a weight, which can be modified [17].

Each neuron in a layer has the same number of inputs, which is equal to the total number of inputs to that layer. Therefore, every layer of p neurons has a total of n inputs and a total of p outputs (each neuron has exactly one output). This implies that each neuron has n inputs as well. In addition, all the neurons in a layer have the same nonlinear transfer function. Each neuron has a number of weights, each corresponding to an input, along with a threshold or bias value. The list of weights can be thought as a vector of weights arranged from 1 to n, the number of inputs for that neuron. Since each neuron in a layer has the same number of inputs, but a different set of weights, the weight vectors from p neurons can be appended to form a weight matrix W of size (n, p).Similarly, the single bias constant from each neuron is appended into a vector of length p. Note that *every* neuron in the net is identical in structure. This fact suggests that the neural net is inherently massively parallel [18].

1.1 Artificial Neural Network Cell (Perceptron)

Artificial neural network consisting of a single layer and a simple activation function is called perceptron. The first formal definition of a synthetic neuron model based on the highly simplified consideration of the biological model was formulated by McCulloch and Pitts [17]. A basic artificial neural network neuron has more simple structure than a biological neuron cell. Biological neuron cell and basic neuron model are shown in Figure 1.1.

The scalar input p is multiplied by the scalar weight w to form w_p, one of the terms that is sent to the summer. The other input, 1, is multiplied by a bias b and then passed to the summer. The summer output, n, often referred to as the net input, goes into a transfer function or activation function f, which produces the scalar neuron, output a [19].

M. Ünal et al.: Optimization of PID Controllers Using AC and GA, SCI 449, pp. 5–17.
springerlink.com

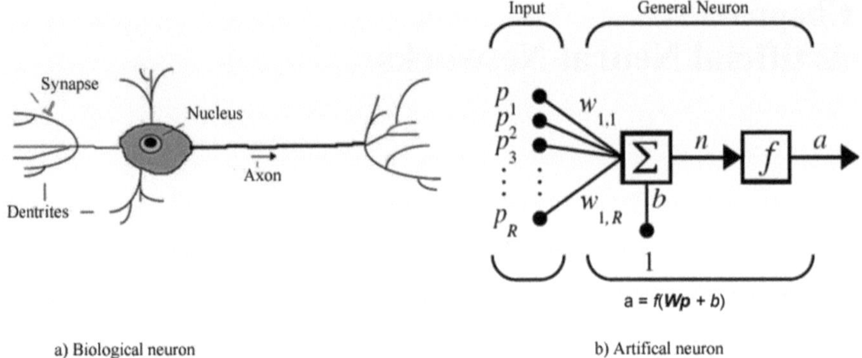

Fig. 1.1 Biological and Artificial Neurons

Typical activation functions used are

$$f(net) \triangleq \frac{2}{1+\exp(-\lambda net)} - 1 \tag{1.1a}$$

and

$$f(net) \triangleq \mathrm{sgn}(net) = \begin{cases} +1, & net > 0 \\ -1, & net < 0 \end{cases} \tag{1.1b}$$

where $\lambda > 0$ in (1.1a) is proportional to the neuron gain determining steepness of the continuous function *f(net)* near $net = 0$. The continuous activation function is shown in Figure 1.2a for various. As $\lambda \to \infty$ the limit of the continuous function becomes the *sgn (net)* function. Activation functions Equation 1.1a and 1.1b are called bipolar continuous and bipolar binary functions, respectively. The word "bipolar" is used to point out that both positive and negative responses of neurons are produced for this definition of the activation function.

By shifting and scaling the bipolar activation functions defined by Eq.1.1, unipolar continuous and unipolar binary activation functions can be obtained, respectively, as follows

$$f(net) \triangleq \frac{1}{1+\exp(-\lambda net)} \tag{1.2a}$$

and

$$f(net) \triangleq \mathrm{sgn}(net) = \begin{cases} 1, & net > 0 \\ 1, & net < 0 \end{cases} \tag{1.2b}$$

Function 2a is shown in Figure 1.2a. The unipolar binary function is the limit of *f(net)* in Equation 1.2a when $\lambda \to \infty$. The soft-limiting activation functions Equation 1.1a and 1.2a are often called sigmodial characteristics, as opposed to the hard-limiting activation functions given in Equation 1.1b and 1.2b [17].

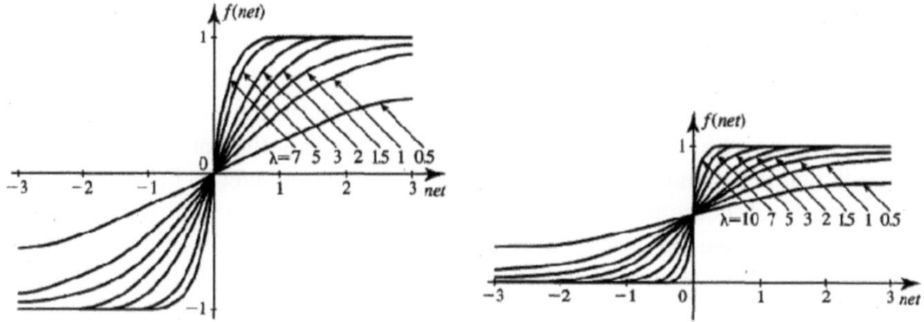

Fig. 1.2 Activation functions of a neuron (a) bipolar continuous (b) unipolar continuous

1.2   Artificial Neural Network Models

ANN models can be examined in two groups according to the structure: Feedforward and Recurrent networks.

1.2.1   Feedforward Network Models

The simplest and most common artificial neural networks use one-way signal flow. There are no delays in the feed forward artificial neural networks, the process advances to the inputs to outputs. Output values are compared with the desired output value and weights are updated with obtained error signal. Feedforward network model is shown in Figure 1.3.

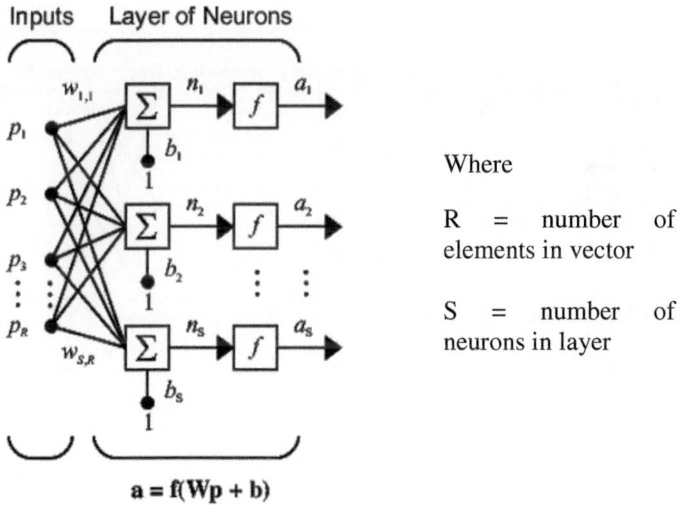

Fig. 1.3 Layer of S neurons

In this network, each element of the input vector p is connected to each neuron input through the weight matrix W. The ith neuron has a summer that gathers its weighted inputs and bias to form its own scalar output $n(i)$. The various $n(i)$ taken together form an S-element net input vector n. Finally, the neuron layer outputs form a column vector a. The expression for a is shown at the bottom of the figure.

Note that it is common for the number of inputs to a layer to be different from the number of neurons (i.e., R is not necessarily equal to S). A layer is not constrained to have the number of its inputs equal to the number of its neurons.

The input vector elements enter the network through the weight matrix W.

$$W = \begin{bmatrix} w_{1,1} & w_{1,2} & \cdots & w_{1,R} \\ w_{2,1} & w_{2,2} & \cdots & w_{1,1} \\ \cdot & \cdot & & \cdot \\ w_{S,1} & w_{S,2} & \cdots & w_{S,R} \end{bmatrix} \tag{1.3}$$

Note that the row indices on the elements of matrix W indicate the destination neuron of the weight, and the column indices indicate which source is the input for that weight. Thus, the indices in $w_{1,2}$ say that the strength of the signal from the second input element to the first (and only) neuron is $w_{1,2}$.

A network can have several layers. Each layer has a weight matrix W, a bias vector b, and an output vector a. To distinguish between the weight matrices, output vectors, etc., for each of these layers in the figures, the number of the layer is appended as a superscript to the variable of interest. A three-layer network is shown in Figure 1.4.

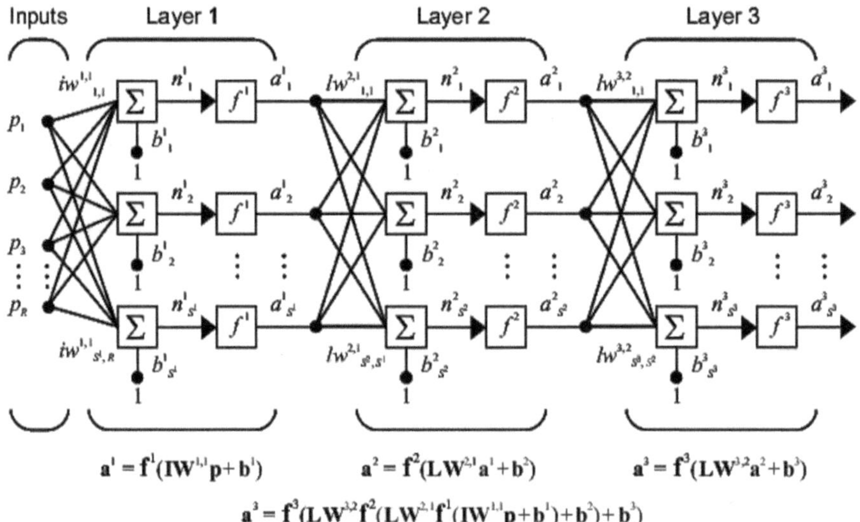

$$\mathbf{a}^1 = \mathbf{f}^1(\mathbf{IW}^{1,1}\mathbf{p} + \mathbf{b}^1) \qquad \mathbf{a}^2 = \mathbf{f}^2(\mathbf{LW}^{2,1}\mathbf{a}^1 + \mathbf{b}^2) \qquad \mathbf{a}^3 = \mathbf{f}^3(\mathbf{LW}^{3,2}\mathbf{a}^2 + \mathbf{b}^3)$$

$$\mathbf{a}^3 = \mathbf{f}^3(\mathbf{LW}^{3,2}\mathbf{f}^2(\mathbf{LW}^{2,1}\mathbf{f}^1(\mathbf{IW}^{1,1}\mathbf{p} + \mathbf{b}^1) + \mathbf{b}^2) + \mathbf{b}^3)$$

Fig. 1.4 Three layer network

The network shown above has R inputs, S^1 neurons in the first layer, S^2 neurons in the second layer, etc. It is common for different layers to have different numbers of neurons. A constant input 1 is fed to the bias for each neuron.

Note that the outputs of each intermediate layer are the inputs to the following layer. Thus layer 2 can be analyzed as a one-layer network with $R = S^1$ inputs, $S = S^2$ neurons, and a $S^1 x S^2$ weight matrix W^2. The input to layer 2 is a^1; the output is a^2. Now, all the vectors and matrices of layer 2 have been identified, it can be treated as a single-layer network on its own. This approach can be taken with any layer of the network.

The layers of a multilayer network play different roles. A layer that produces the network output is called an output layer. All other layers are called hidden layers.

1.2.2 Recurrent Networks

A feedback network can be obtained from the feedforward network by connecting the neurons' outputs to their inputs. There are delays in artificial neural networks with feedback. One type of discrete-time recurrent network is shown in Figure 1.5.

D shows the time delay. D holds the information at the past sample time in memory and it is applied to the next output as input.

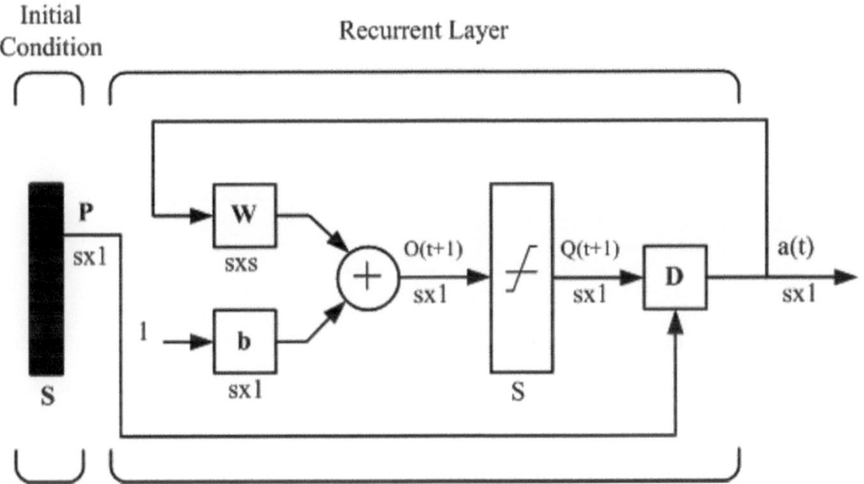

Fig. 1.5 Recurrent Network

1.2.3 Learning Rules

A learning rule is defined as a procedure for modifying the weights and biases of a network (This procedure can also be referred to as a training algorithm). The learning rule is applied to train the network to perform some particular tasks. Learning rules fall into three broad categories: supervised learning, unsupervised learning and reinforcement (or graded) learning. The block diagrams of supervised learning and unsupervised learning are illustrated in Figure 1.6 [17].

In supervised learning we assume that at each instant of time when the input is applied, the desired response d of the system is provided by the teacher. This is illustrated in Figure 6(a). The distance *p[d, o]* between the actual and the desired response serves as an error measure and is used to correct network parameters externally. Since we assume adjustable weights, the teacher may implement a reward-and-punishment scheme to adapt the network's weight matrix W. This mode of learning is very pervasive. Also, it is used in many situations of natural learning. A set of input and output patterns called a training set is required for this learning mode. Typically, supervised learning rewards accurate classifications or associations and punishes those which yield inaccurate responses. The teacher estimates the negative error gradient direction and reduces the error accordingly. In many situations, the inputs, outputs and the computed gradient are deterministic, however, the minimization of error proceeds over all its random realizations. As a result, most supervised learning algorithms reduce to stochastic minimization of error in multi-dimensional weight space [17].

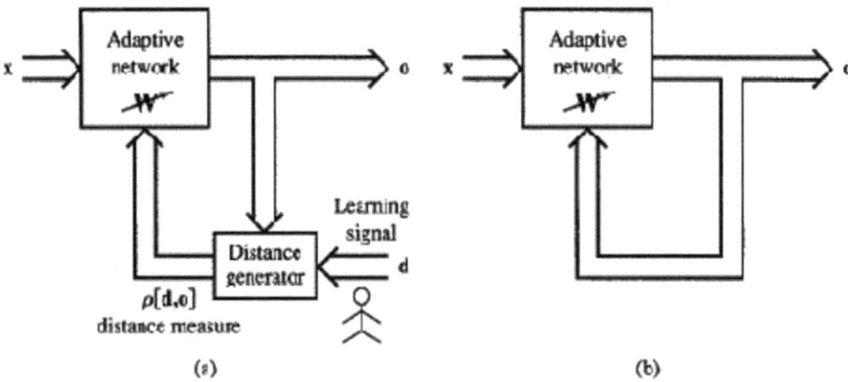

Fig. 1.6 Block diagram for explanation of basic learning modes: (a) Supervised learning and (b) Unsupervised learning

Figure 6(b) shows the block diagram of unsupervised learning. In learning without supervision, the desired response is not known; thus, explicit error information cannot be used to improve network behavior. Since no information is available as to correctness or incorrectness of responses, learning must somehow be accomplished based on observations of responses to inputs that we have marginal or no knowledge about. Unsupervised learning algorithms use patterns that are typically redundant raw data having no labels regarding their class membership, or associations. In this mode of learning, the network must discover for itself any possibly existing patterns, regularities, separating properties, etc. While discovering these, the network undergoes change of its parameters, which is called self-organization. The technique of unsupervised learning is often used to perform clustering as the unsupervised classification of objects without providing information about the actual classes. This kind of learning corresponds to minimal a priori information available. Some information about the number of clusters, or similarity versus dissimilarity of patterns, can be helpful for this mode of learning.

Unsupervised learning is sometimes called learning without a teacher. This terminology is not the most appropriate, because learning without a teacher is not possible at all. Although, the teacher does not have to be involved in every training step, he has to set goals even in an unsupervised learning mode [20]. It may think of the following analogy. Learning with supervision corresponds to classroom learning with the teacher's questions answered by students and corrected, if needed, by the teacher. Learning without supervision corresponds to learning the subject from a videotape lecture covering the material but not including any other teacher's involvement. The teacher lectures directions and methods, but is not available. Therefore, the student cannot get explanations of unclear questions, check answers and become fully informed [17].

Reinforcement learning is a learning method for agents to acquire the optimal policy autonomously from the evaluation of their behavior. The indication of the performance is called the reinforcement signal. In this learning scheme, the learning agent receives sensory input from the environment. Then it decides and carries out an action based on the input. In return, it receives a reinforcement signal, and new sensory input. The agent uses these episodes to maximize the reinforcement signals aggregated in time [21].

1.2.4 Multilayer Perceptron

Multilayer Perceptron (MLP) consists of three-layers of neurons (input, hidden and output layer as shown in Figure 1.7) interconnected by weights. Neural networks are those with at least two layers of neurons-a hidden and an output layer, provided that the hidden layer neurons have nonlinear and differentiable activation functions. The nonlinear activation functions in a hidden layer enable a neural network to be a universal approximator. Thus, the nonlinearity of the activation functions solves the problem of representation. The differentiability of the hidden layer neurons activation functions solves the nonlinear learning task. Generally, no processing will occur in the input layer.

The input layer is not treated as a layer of neural processing units. The input units are merely fan-out nodes. Generally no processing will occur in the input layer, and although in it looks like a layer, it's not a layer of neurons. Rather, it is an input vector, eventually augmented with a bias term, whose components will be fed to the next (hidden or output) layer of neural processing units. The output layer neurons may be linear or they can be having sigmoidal activation functions [22].

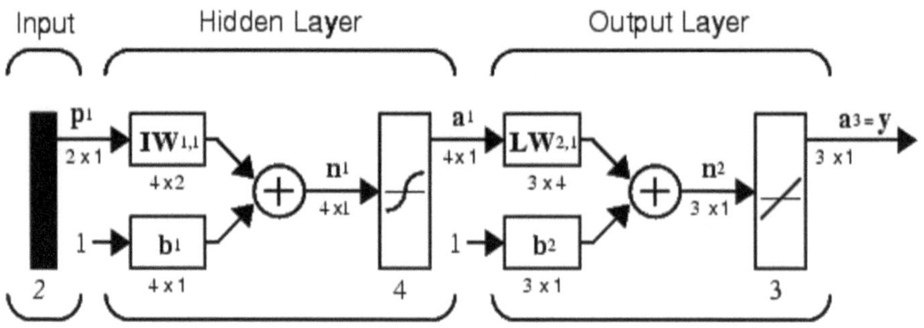

Fig. 1.7 Multi Layer Perceptron

1.2.5 Error Backpropagation Algorithm

The basic idea behind the error backpropagation algorithm is that the error signal terms for hidden layer neurons are calculated by backpropagation the error signal terms of the output layer neurons. Backpropagation is still the most commonly used learning algorithm in the field of soft computing.

MLP network starts with random initial values for its weights, and then computes a one-pass backpropagation algorithm at each time step k, which consists of a forward pass propagating the input vector through the network layer by layer, and a backward pass to update the weights by the gradient descent rule. After having trained on-line for a period of time, the training error should have converged to a value so small that if training were to stop, and the weights frozen. Also, the training of NNs is said to have reached a global minimum when, after changing the operating conditions, as well as freezing the weights, the network's response is still reasonably acceptable [20]. Steps of the algorithm are as following:

Initialize connection weights into small random values.

Present the *p*th sample input vector of pattern $X_p = (X_{p1}, X_{p2}, ..., X_{pN_0})$ and the corresponding output target $T_p = (T_{p1}, T_{p2}, ..., T_{pN_M})$ to the network.

Pass the input values to the first layer, layer 1. For every input node *i* in layer 0, perform:

$$Y_{0i} = X_{pi} \tag{1.4}$$

For every neuron *i* in every layer j = 1,2, ... M, from input to output layer, find the output from the neuron:

$$Y_{ji} = f\left(\sum_{k=1}^{N_{j-1}} Y_{(j-1)k} W_{jik}\right) \tag{1.5}$$

where *f* is the activation function. The most widely used activation function for neurons in the hidden layer is the following sigmoidal function:

$$f(x) = \frac{1}{1 + e^{-x}} \tag{1.6}$$

Obtain output values. For every output node i in layer M, perform:

$$O_{pi} = Y_{Mi} \tag{1.7}$$

Calculate error value δ_{ji} for every neuron i in every layer in backward order $j = M, M-1, \ldots .2,1$ from output to input layer, followed by weight adjustments. For the output layer, the error value is:

$$\delta_{Mi} = Y_{Mi}(1 - Y_{Mi})(T_{pi} - Y_{Mi}) \tag{1.8}$$

and for hidden layers:

$$\delta_{Mi} = Y_{ji}(1 - Y_{ji}) \sum_{k=1}^{N_{j+1}} \delta_{(j+1)k} W_{(j+1)ki} \tag{1.9}$$

The weight adjustment can be done for every connection from neuron k in layer (i-1) to every neuron i in every layer i:

$$W_{jik}^{+} = W_{jik+}\beta\delta_{ji}Y_{ji} \tag{1.10}$$

where β represents weight adjustment factor normalized between 0 and 1.

The actions in steps 2 through 6 will be repeated for every training sample pattern p, and repeated for these sets until the Root Mean Square (RMS) of output errors is minimized.

RMS of the errors in the output layer defined as:

$$E_p = \frac{1}{2} \sum_{j=1}^{N_m} \left(T_{pj} - O_{pj}\right)^2 \tag{1.11}$$

for the pth sample pattern [21].

1.2.6 *The Nonlinear Autoregressive Network with Exogenous Inputs (NARX) Type ANN*

A simple way to introduce dynamics into network consists of using an input vector composed of past values of the system inputs and outputs. This the way by which the MLP can be interpreted as the nonlinear autoregressive network with exogenous inputs (NARX) model of the system. This way of introducing dynamics into a static network has the advantage of being simple to implement [23].

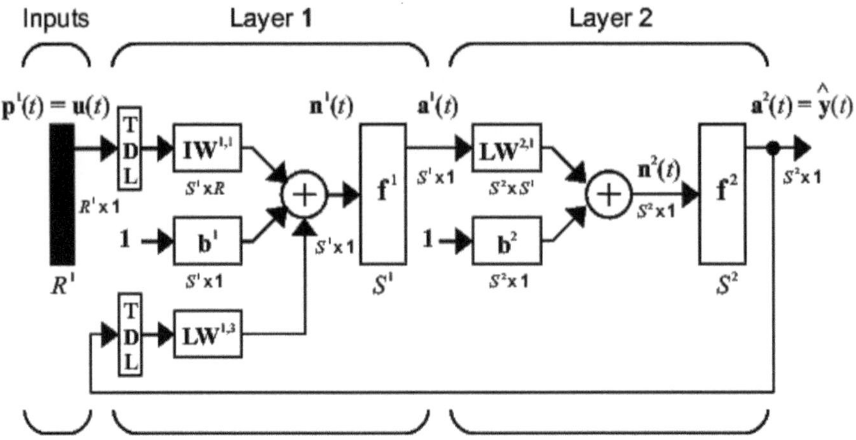

Fig. 1.8 NARX model

NARX model can be implemented by using a feedforward neural network to approximate the function f. A diagram of the resulting network is shown below, where a two-layer feedforward network is used for the approximation.

The defining equation for the NARX model is

$$\hat{y}(k) = f(y(k-1),...,y(k-n);u(k-1),...,u(k-m))+\varepsilon(k) \tag{1.12}$$

where $\hat{y}(k)$ is predicted value, $y(k)$ process output, $u(k)$ process input, $\varepsilon(k)$ is the approximation error at time instant k, m is an input time delay, n is an output time delay, f is a non-linear function (activation function) describing the system behavior.

There are many applications for the NARX network. It can be used as a predictor, to predict the next value of the input signal. It can also be used for nonlinear filtering, in which the target output is a noise-free version of the input signal. The use of the NARX network is demonstrated in another important application, the modeling of nonlinear dynamic systems.

It can be considered the output of the NARX network to be an estimate of the output of some nonlinear dynamic system. The output is fed back to the input of the feedforward neural network as part of the standard NARX architecture, as shown in the left Figure 1.9. Because the true output is available during the training of the network, it can be created a series-parallel architecture, in which the true output is used instead of feeding back the estimated output, as shown in the right Figure 1.9. This has two advantages. The first is that the input to the feedforward network is more accurate. The second is that the resulting network has a purely feedforward architecture, and static backpropagation can be used for training.

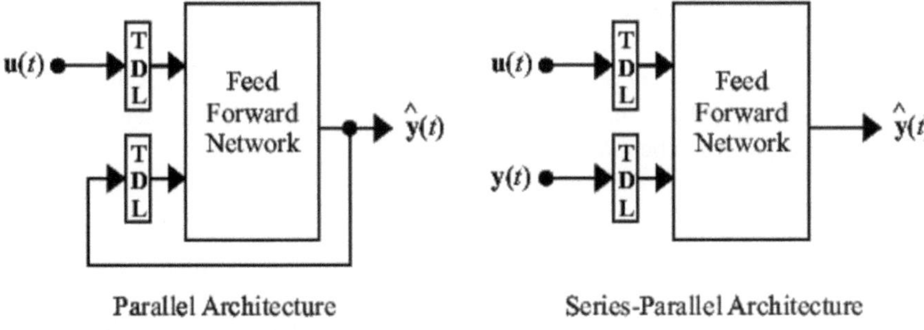

Fig. 1.9 A series-parallel architecture [19]

1.3 Modeling of a System with ANN

Modeling and measurement has very great significance on systems design and control. All elements and link styles of systems must have analytical expressions based on physical laws in classic modeling concept. In fact, this process is not possible or very difficult in complex system. So, in the modern concept of modeling, an appropriate model is proposed to the system and this model is taught with the help of input-output data set and updated. In recent years, especially in complex and nonlinear systems modeling, ANN has been used widely. ANN's nonlinear structure provides a good fit with systems to be taught.

In general, obtaining the system model means that learning of the physical structure of the system based measurement of input, output and state variables. One or more state variables often can be system output. A system model can be specified using the mathematical expression, input-output data set or linguistic rules. ANN can be used for modeling linear and nonlinear systems.

1.3.1 Advanced Modeling (Series-Parallel Modeling)

The input and output changing over time of a dynamic system are shown as, u and y_p respectively. If the model and the system are running the same system input u, $(\hat{y})$'s output produced by the model should be closer to y_p. Block diagram of this method is shown in Figure 1.10.

Identification Artificial Neural Network (ANN) blocks in Figure 1.10, is placed in parallel with the non-linear system. The error between system output y_p and ANN output $\hat{y}(k)$, $e(k) = y_p - \hat{y}$, is used in training of ANN. It's assumed that system is described by the following non-linear discrete-time difference equation.

$$y_p(k+1) = F\left[y_p(k),...,y_p(k-n+1);u(k),...,u(k-m+1)\right] \qquad (1.13)$$

Here, $y_p(k+1)$ is the value at the $(k+1)$ sampling time of system output, $F(.)$ is multi-dimensional system definition dependent to past system output (n) and past input (m) where n and m are positive integers. It's assumed that ANN 's input-output structure is the same with system in input-output structure selection [24]. So, ANN output's can be indicated as follows:

$$\hat{y}(k+1) = y_p^{'}(k+1) = N\left[y_p(k),...,y_p(k-n+1);u(k),...,u(k-m+1)\right] \qquad (1.14)$$

Here, $N(.)$ is nonlinear definition of the ANN and $\hat{y}$ is the predicted ANN output. As can be seen here, ANN output depends on the past n system output and m system input.

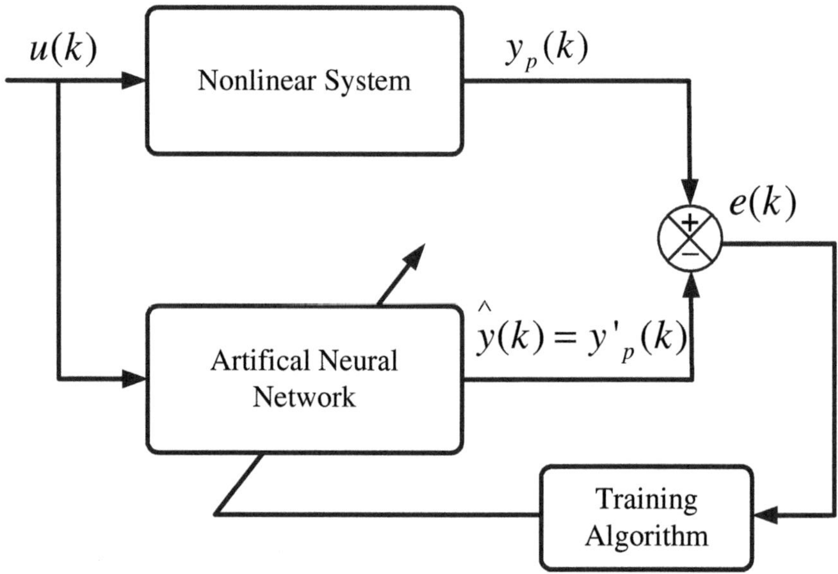

Fig. 1.10 The basic advanced type block diagram using on modeling with ANN

Figure 1.11 shows the block diagram of the identification algorithm created by using Equation 1.13 and 1.14. In this serial-parallel identification system, the main principle is to provide the system output and its backward modes as input to ANN model. Time Delay Layer (TDL) blocks in Figure 1.11 refer cycles of time delay [24-26].

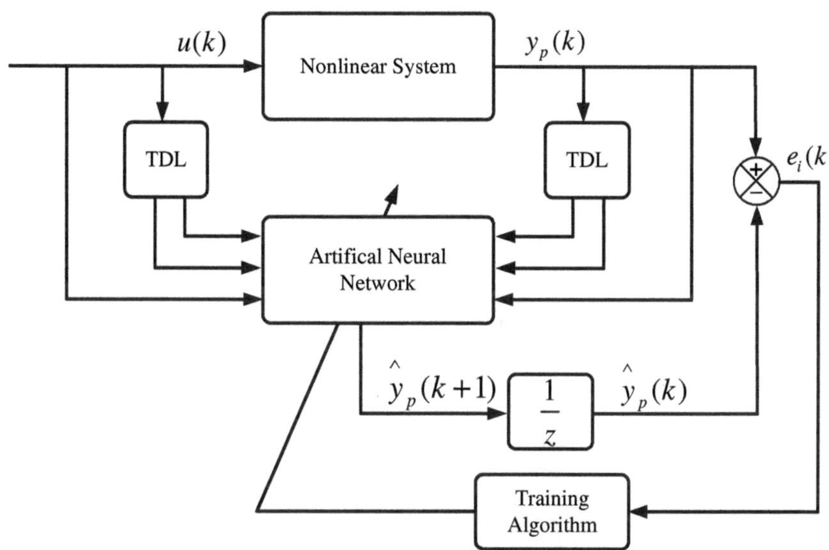

Fig. 1.11 Block diagram of serial – parallel advance type modeling

Chapter 2
Genetic Algorithm

Evolutionary algorithms (EAs) are global, parallel, search and optimization methods, founded on the principles of natural selection and population genetics. In general, any iterative, population based approach that uses selection and random variation to generate new solutions can be regarded as an EA. The evolutionary algorithms field has its origins in four landmark evolutionary approaches: evolutionary programming (EP), evolution strategies (ES), genetic algorithms (GA), and genetic programming (GP). The genetic algorithm was popularized by Goldberg (1989) and, as a result, the majority of control applications in the literature adopt this approach.

GAs has been used to solve a wide range of optimization problems during the last thirty years. Their applications include several types of problems such as designing the communication networks, optimizing the database query, and controlling the physical systems. Thus, GA has become a robust optimization tool for solving the problems related to different field of the technical and social sciences [27, 28].

GA is the process of searching the most suitable one of the chromosomes that built the population in the potential solutions space. A search like this, tries to balance two opposite objectives: searching the best solutions (exploit), and expanding the search space (explore) [29].

GA starts parallel searching from independent points of search space in which the solution knowledge is poor or not available. The solution depends on interaction of the surroundings and genetic operators. For that reason, obtaining the suboptimal solutions of GA is a small probability. It is particularly presented as an alternative for the traditional optimal search approaching in which it is hard to find the global optimum point in nonlinear and multimodal optimizations problems. GA provides solutions using randomly generated bit strings for different types of problems. Any GA could be defined by 11 parameters given as below:

$$\Sigma, G, \rho, f, \mu, I, S, Gap, \Omega, R, \tau \tag{2.1}$$

where, Σ represents the search space (phenotype), G is the coding space (genotype), ρ is the coding function ($\rho:\Sigma{\rightarrow}G$), f is the fitness function ($f:\Sigma{\rightarrow}R^+$), μ is the population size, I is the initialization function ($I:\mu{\rightarrow}P(G)$), $P(G)$ is the population, S is the selection type, Gap is the percentage of the population that conveyed from old generation to new generation, Ω is the genetic operator

M. Ünal et al.: Optimization of PID Controllers Using AC and GA, SCI 449, pp. 19–29.
springerlink.com

(crossover operator: ω_c, and mutation operator: ω_m), R is the replacement policy, and τ is the termination criterion.

A simple GA structure, which is built on these parameters, is given in Figure 2.1 [30]. Selection of the parameters depends on the type of the problem. In literature, there are a lot of studies to find the best fit of the parameters. However, the way to best fit the parameters, which can be used for all the problem types [30-33].

```
t := 0;
ρ:Σ→G
FOR i := 1  TO  μ DO I : μ → P(G)  [ Initial population(P₀, ( p₀, p₁ , p_μ)) ];
FOR i := 1  TO  μ DO f:Σ→R+ [ Fitness ( pᵢ ) ];
  WHILE   τ ( termination condition not fulfilled )  DO
     BEGIN
t := t+1
FOR i := 1  TO  μ DO S[ Select an individual ( ( p_{i,t} ) from P_{t-1} ) ];
FOR i := 1  TO  μ −1  STEP  2  DO
     IF Random  ([0, 1] ≤ P_c) THEN ω_c [ Cross  ( p_{i,t} , p_{i+1,t} ) ] ;
FOR i := 1  TO  μ  DO
IF Random  ([0, 1] ≤ P_m)  THEN  ω_m [ Mutate  ( p_{i,t} ) ] ;
FOR i := 1  TO  μ  DO  f : Σ → R+[ Fitness  ( p_{i, t+1} ) ];
FOR i := 1  TO μ DO Gap [ P_{i,t+1} := P_{i,t} ];
     END; END.
```

Fig. 2.1 Simple GA structure

There are three types of GA models commonly used: Simple GA (SGA), micro GA (μGA), and Steady State GA (SSGA). An example of the SGA model is described in Fig.1. It can be used for many function optimization problems. However, it has a disadvantage that it takes a long time to evaluate the large populations. Using the small population μGA reaches to the optimal zone faster. To achieve success using the small populations, new chromosomes are produced with proper intervals. In the SSGA models a certain population is transferred from previous generation to the next generation.

The basic difference between the GA and other research methods is that GA uses coded parameters instead of the parameters themselves. Coding method has a grand effect on the performance of the GA. So, the first step of the GA application is to select the most appropriate coding, G, which will represent the solution space of the problem. Generally, GA uses binary coding. However, it can use various other methods like reel number coding, tree coding or symbolic alphabets coding [29, 32].

GA starts to search from more than one point in the search space, Σ, to find the optimal solution. Usually the first population is created coincidentally, representing the whole search space. Selection of the accommodate population size, μ, is another important decision for GA applications. If the size of the population is too small, mostly because of the insufficient information, it may result to a suboptimal solution [31, 33-35].

Replacement, *Gap,* is the operation in which the individuals of the new generation are selected from parents. Frequently used replacement methods are the *generation gap* and the *steady-state* methods. The steady-state replacement method is used in SSGA method. Only one individual is produced for each generation [35-37].

The principal aim of the selection, *S,* is selecting the best chromosomes to increase the individual fitness value. There are three important selection methods: the stochastic sampling with replacement selection (SSRS) (or roulette wheel) method, the stochastic universal sampling (SUS) method, and the tournament selection (TS) method [34].

The crossover operator, ω_c, can be described as exchanging two chromosomes chosen as the parents. Traditionally, crossover point number is one. However, there are some essays suggesting more crossover points and mutation proportion. The other important crossover method is *uniform crossover*. In this method, each parent's genes crossed over with different probabilities [38].

Mutation operator, ω_m, changes one or more genes in the chromosome. Traditionally, mutation operator is used rarely [31]. Usually, the crossover probability (P_c) is chosen between 0.25 and 1, and mutation probability (P_m) is chosen between 0.01 and 0.001. Some studies use dynamic mutation and crossover probabilities [39].

2.1 Types of Genetic Algorithms

In the literature this structure is described as Simple Genetic Algorithm SGA. In many algorithm applications, this SGA model is used. SGA forms a reference for other sequential algorithm types. When we look at the literature, in the sequential algorithm types, some other models are also described, Micro Genetic Algorithm (μGA), steady state Genetic Algorithm (ssGA), Hierarchic Genetic Algorithm (HGA) and Messy Genetic Algorithm (mGA).

2.1.1 Micro Genetic Algorithm (μGA)

SGA is known as useful for many function optimization problems. But the disadvantage of SGA is that it takes long time to evaluate the coherence of the generation for large populations. This is why, small population approach which is constituted with simple genetic parameters is proposed [30, 32]. Using small population size μGA reaches faster to optimal zone than. μGA, uses double coded chromosome structure. Population size of μGA is constant 5. To achieve success using the small populations, new chromosomes are produced in the population with proper intervals. The aim of μGA is not increasing the value of adaptation of population but it is achieving the best solution faster.

2.1.2 Steady-State Genetic Algorithm

In Steady state GA; a certain population is transferred from previous generation to the next. The percentage of this certain population is determined by the user and based on their value of fitness. But this value remains constant through generations and so operation time is saved [37].

2.1.3 Hierarchic Genetic Algorithm (HGA)

When we analyze the biological structure of DNA, we can see that the structure forms a hierarchic model. In this hierarchic structure some genes task is to determine the active and inactive genes. HGA algorithm tries modeling this hierarchic structure of biologic DNA[40]. In this chromosome structure, the values of control genes determine which parameter Gene will be active or not. If the HGA algorithm is formed this way, all the operators used in SGA can be adapted. But the crossover and mutation operators adapted to control gene structure will affect parameter gene structure. Consequently it will make a second effect on the chromosome.

2.1.4 Messy Genetic Algorithm (mGA)

In mGA structure variant length of chromosome is used while in the SGA structure the length of chromosome is constant. Since in the mGA algorithm length of the chromosome varies, "cut and splice" operator is used instead of "crossover "operator which is used in SGA. Chromosomes which have different lengths are obtained. The other genetic operators used in mGA are same as the ones used in SGA [41].

2.2 Scheme Theorem and Genetic Algorithm Operators

Scheme theorem is based on the resemblances between chromosomes [29]. The equation given by Holland (Equation 2.2) is described as "equality of development" and shows the probability of finding appropriate offspring in the next generations.

$$\xi_{(S,t+1)} \geq \xi_{(S,t)} \frac{\Phi_{(S,t)}}{\Phi_{(t)}} \left[1 - Pc\frac{\delta_{(S)}}{L-1} - O_{(S)}Pm \right] \tag{2.2}$$

where Pm is mutation probability, *Pc* is crossing probability, $O_{(S)}$ is grade of scheme, *L* is length of the bit string, $\delta_{(S)}$ is the length of the scheme, $\xi_{(S,t)}$ is the number of series matching with scheme *S* in t time, $\Phi_{(S,t)}$ is fitness of *S* scheme in t time, $\Phi_{(t)}$ is approximate value of fitness.

2.2.1 Coding (G)

The basic difference of genetic algorithm from other research methods is that genetic algorithm uses coded parameters instead of parameters themselves. This is why; the first step of the genetic algorithm application is to select the most appropriate coding method which will represent the solution space of the problem better. Generally; genetic algorithm uses binary coding method. In the literature there are different genetic algorithm types using integer, reel number or three schemes, and also there are certain algorithm studies which use different symbolic alphabets [29, 36].

$$x_{\text{Real}} := x_{\text{min}} + \frac{x_{\text{max}} - x_{\text{min}}}{2^L - 1} \left(\sum_{i=0}^{L-1} b[i] * 2^{L-i-1} \right) \tag{2.3}$$

Coding method has a grand effect on the performance of the genetic algorithm, but it is impossible to name the most appropriate method since coding method depends on the problem itself. Michalewicz shows that using reel numbers results faster but for a certain problem. In some studies gray coding method gives better results than binary coding.

2.2.2 Creating First Population (Σ)

Genetic algorithm starts searching from more than one point to find the optimal resolution. So, it is important to create the initial values of these points. Usually first population is coincidentally created as representing the whole search space. But some studies; especially in limited optimization problems, shows that first population is produced depending on a prior information or heuristically [42].

2.2.3 Size of the Population (μ)

Selection of the accommodate population size is an important decision of genetic algorithm application. Some studies tried to find valid criterions for population size [34, 37, 43, 44]. If the size of the population is too small, mostly because of the insufficient information, it will result to an in- optimal solution. Because, it brings a limited solution and it converges quickly. On the other hand, a too large size of the population values takes too much time to result.

The theoretical base to determine the size of the population is, Holland's $Q(n^3)$ estimation of schemas [30].

According to this scheme; the number of schemes operated by a genetic algorithm is proportional with the cube of the size of the population. Grefenstette, claims that the best population size for genetic algorithm should be between 10 and 160 [45]. But he also shows that there are nonlinear relations between crossover and mutations probabilities and the population size. Odeyato has suggested a size between 100 and 400; and Robertson used size till 8000 for classification problems [33]. Another study was realized by Goldberg who analyses the appropriate population size for consecutive and parallel genetic algorithms. He found a value which can compare population sizes and convergence times. This value can be shown as [34]:

$$\mu = (O_o - O_c)/T_c \tag{2.4}$$

where, O_o is the initial number of population scheme, O_c is the number of population schemes which are converging, T_C is the time until convergence.

2.2.4 Reproduction (Gap)

Reproduction is the operation in which the offsprings who will form the new generation are selected from parents. Reproduction is an operator which produces identical offspring with hits parents. Frequently used reproduction techniques are complete replacement, generation gap and steady state methods [30, 32].

- **Complete Replacement**

  This technique contains the replacement of all the offsprings of ex-generation with the new generation offsprings. The biggest disadvantage of this method is the risk of loss of offsprings which has the appropriate values. To eliminate this disadvantage, elistic models are used. Elistic models transfer the appropriate offsprings to the next generation.

- **Generation Gap**

  This method resembles with the complete replacement method. The difference between two methods is, in generation gap a certain percentage of the offsprings are transferred to the new generation. For example, 0.5 generation gap means that the 50% of the offsprings of ex-generation are transferred to the next generation. A generation gap of 100 % forms complete replacement method. Offsprings which will be transferred can be selected randomly or an elistic model can be used. But generation gap method cannot guarantee better fitness values for the new generation offsprings than the values of their parents. So, especially when a small generation gap is chosen, it will give completely different result than complete replacement method.

- **Steady State**

  Steady State reproduction method is the reproduction method which is used in SSGA. This method contains erasing only one offspring form the old generation and reproduces only one offspring. So for each generation only 1 offspring is produced.

2.2.5 Selection (S)

The essential principal of the selection is generally "the best survives". The aim of the selection is to increase the offspring which have better fitness values. Selection methods can be grouped to 3 [7, 20].

- **Fitness Proportionate Selection**

Checking the fitness values of parents to form the offsprings and to select the parents is the general characteristic of these methods. Generally, these methods are Sampling with Replacement (Roulette Wheel) and Stochastic Universal Sampling (SUS).

o *Stochastic Sampling with Replacement Selection:*

In this method expected selection probability of the offspring is determined by the division of the value of fitness of the offspring to total fitness of the generation. Through this method, sometimes probability of best offspring's loss occurs. To eliminate this probability, scaling and/or using elistic models may be necessary.

This method is generally compared to roulette. Total value of fitness of the offspring in the generation gives the roulette wheel and each offspring hold a place equal to its fitness value. In the Roulette wheel selection method the probability of an offspring to be selected is given in Equation (2.5) [46].

$$p_i = \frac{\Phi_i}{\sum_{j=1}^{\mu} \Phi_j} \tag{2.5}$$

where p_i is the i (probability of selection).

o *Stochastic Universal Sampling (SUS):*

This selection method is developed to cover the difference between the probability of selection in the roulette wheel method and eventual selection value. In SUS a roulette wheel according to fitness value of the society is created such as the anterior method. To select the offsprings, wheel is turned once instead of turning it as the size of the population. But between the roulette signs, there are equal distances and the number of the sings is equal to the population size. Offsprings shown by the roulette signs are selected as parents for the next generation. Algorithm of SUS selection method is given Figure 2.2.

```
FOR i=1 TO μ DO
BEGIN
      x := Random[0, 1];
      k := 1
```

$$\textbf{WHILE } k < \mu \ \& \ x < \sum_{j=1}^{k} f(p_{j,t}) / \sum_{j=1}^{\mu} f(p_{j,t}) \ \textbf{DO}$$

```
k := k + 1;
      p_{i,t+1} := p_{k,t};
END
```

Fig. 2.2 SUS Selection Algorithms

- **Tournament Selection Method**

In this method; offsprings are randomly grouped. In each group, offsprings compete with each other. In each group the offspring which has the highest fitness value is chosen as parent for next generation. Operation continues until the total population is created. In this method group size is important and highly affects the method's performance. For some applications group size is determined as 2, and for some much bigger groups are developed. Tournament selection method is deterministic and gives better results in application in which the groups are smaller. Especially when compared to other proportional selection methods [27]. The probability of selection for an offspring is shown in Equation (2.6) [46].

$$p_i = \frac{1}{\mu^q}((\mu-i+1)^q - (\mu-i)^q) \tag{2.6}$$

where q is the size of the tournament.

2.2.6 *Crossover (ω_c)*

Crossover operation can be described as the exchange of 2 chromosomes chosen as parents. These chromosomes are cut from a point chosen randomly but in a way they will make a whole. Traditionally in genetic algorithm, crossover point number is 1 and scheme theorem is designed according to crossover operation from 1 point. In Figure 2.3 one point crossover algorithm is given. But there are some essays suggesting more crossover points and mutation proportion [29, 32].

```
FOR i:=1 TO μ-1 STEP 2 DO
BEGIN
    IF Random [0,1] ≤ Pc THEN
    BEGIN
pos:=Random (1,......L-1)
        FOR i:=1 TO pos DO
        BEGIN
Offspring₁[i]:=Parent₁[i];
Offspring₂[i]:=Parent₂[i];
        END
        FOR i:=pos+1 TO L DO
        BEGIN
Offspring₁[i]:=Parent₂[i];
Offspring₂[i]:=Parent₁[i];
        END
    END
END
```

Fig. 2.3 Single Point Crossover Algorithm

- **k- Point Crossover**

In point crossover method, a certain number of k for example) are chosen randomly. To form its own chromosomes; individual takes the genes in segments from parents. In Figure 2.4 crossover from 3 points and new offspring are created from the 4 segments. Crossover probability *(Pc)* shows the probability of the crossing over of the parents in a population to create new individuals. For example; if *Pc=1* all parent will be crossover.

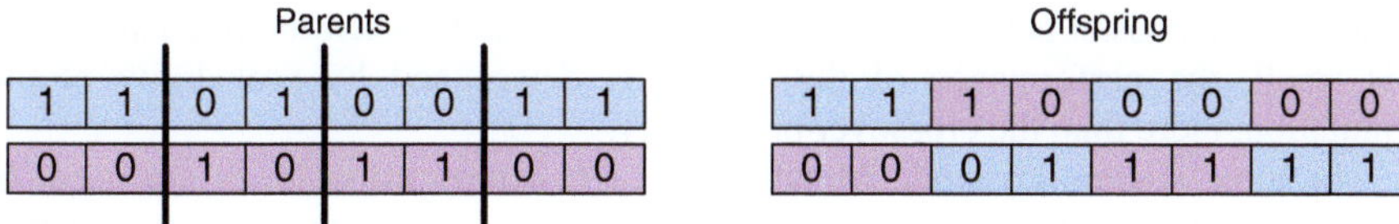

Fig. 2.4 k-point Crossover

- **Crossover String**

In this crossover method; each parent's genes are crossed over with different probabilities. In some applications crossover probabilities of genes are saved on one separate gene [38]. In crossover string, there is a probability *(Pc)* of taking the gene which matches to which parent. For example; if *Pc=* 0.7, offspring will take one parents genes with 0.7 probability; and the genes of another parent with 0.3 probability. If *Pc=0.5* the probability of taking the gene will be equal. For this reason crossover will be random. In Figure 2.5 the structure of crossover string is given.

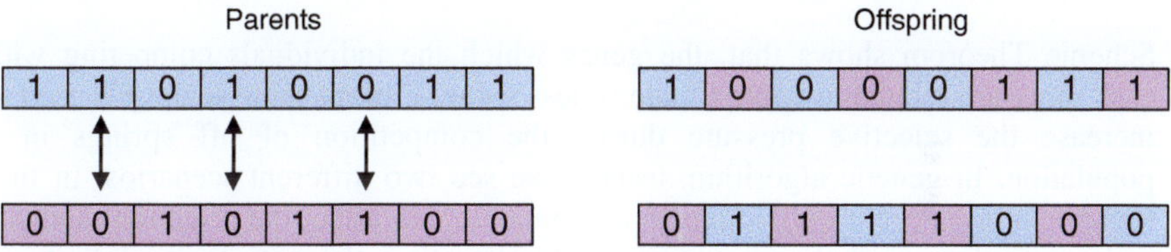

Fig. 2.5 Crossover string

2.2.7 *Mutation (ω_m)*

Mutation is the transformation of one or more genes in a series which creates the offspring. Traditionally; mutation operator is used very rarely. Latest studies show that mutation operator plays a very important role in genetic algorithm. Optimal proportion of probability of mutations *(Pm)* is proportional with the length of the chromosome and the resolution space of the problem.

```
FOR i:=1 TO L DO
    FOR k:=1 TO L DO
IF Random [0,1] < Pm THEN invert pi,t+1[k];
```

Fig. 2.6 Mutation Algorithm

For example; if L is the length of the chromosome, probability of mutation of *k*.degree function must be *Pm= k /L*.

Crossover probability is usually chosen between 0.25 and 1, probability of mutation is chosen between 0.01 and 0.001. Especially, when the population size is small, the performance of the system is determined by probability of the mutation, more than the crossover probability.

In some studies Dynamic Mutation and Crossover probabilities are used. In these studies, the more generation number increases, the probability of mutation also increases but, crossover probability decreases.

Dynamic crossover and mutation probabilities can be expressed as follows [39].

$$P_c = e^{(-T_G / M)} \tag{2.7}$$

$$P_m = e^{(0.005 * T_G / M)} \tag{2.8}$$

Here; T_G shows number of generations in a moment, M shows maximum number of generations.

In some studies, mutation and crossover probability values are heuristically changed according to approximate fitness values of the individuals and the population [47].

2.2.8 Fitness Value Scaling

Scheme Theorem shows that, the genes which the individuals competing with each other are related to approximate fitness value. Fitness value scaling is used to increase the selective pressure during the competition of off springs in a population. In genetic algorithm, usually we see two different scenarios. In first scenario a few decent individuals in a population can affect the selection methods and can enclose a whole population. In this situation, the operation results to an end unpleasantly early. In the second scenario, the individuals in a population start to resemble each other and so the search almost becomes random. There is no reason to prefer an offspring to another one. Fitness value scaling is used to eliminate these two adverse situations [31, 40, 48]. There are three basic methods for scaling. These are:

- **Linear Scaling**

  In this method fitness value is found by using a linear function. The fitness function can be given as follows:

$$\phi' = a * \phi + b \tag{2.9}$$

ϕ shows the fitness value.

Co-efficient a and b are chosen in a way which will ensure the existence of the copies of individuals who has the approximate fitness value and control the fitness value of the individuals who has the maximum value in the present population.

- **Sigma Scaling**

This method is used to get rid of the negative values which occur after linear scaling. The operation is shown below:

$$\phi' = \phi - (\phi_{ort} - c * \sigma) \tag{2.10}$$

In the figure; ϕ shows fitness value, σ; shows standard deviation of the population, c shows the constant which takes a value between 1 and 5.

- **Power Scaling**

In this approach, new fitness value is found by taking the exponent of the existing fitness value. Situation is shown below.

$$\phi' = \phi^k \tag{2.11}$$

Here k depends on the problem and can be changed during the study.

Chapter 3
Ant Colony Optimization (ACO)

The ant colony optimization algorithm (ACO) is an evolutionary meta-heuristic algorithm based on a graph representation that has been applied successfully to solve various hard combinatorial optimization problems. Initially proposed by Marco Dorigo in 1992 in his PhD thesis [49], the main idea of ACO is to model the problem as the search for a minimum cost path in a graph. Artificial ants walk through this graph, looking for good paths. Each ant has a rather simple behavior so that it will typically only find rather poor-quality paths on its own. Better paths are found as the emergent result of the global cooperation among ants in the colony [13, 15, 50-52].

The only way to solve some problems takes place with testing all the possibilities of the answer one by one. The classical example of this is the traveling salesman problem. In this problem is handled to find the shortest route of a businessman circulating city to city. Such problems can be solved by methods applied by moving ants leaving a trace behind them. In 1989 and 1990, in Santa Fe Institute, Bonabeau and his colleagues demonstrated that similar problems can be solved more easily with computers creating "virtual ants".

According to this, virtual ants leave a scent that represents the length of the route on their behind and by this means the other virtual ants will find the shortest routes. Smell of the traces of long routes that are not preferred will disappear gradually by simulating evaporation of the smell trace at a certain speed. This will prevent the deflection of the virtual ants to the long road outside short cut roads [53]. Behavior templates of natural ants which can be used in many optimization problems, especially in the shortest path problems, was introduced by Marco Dorigo and his colleagues for the first time in 1992 [54].

3.1 Real Ants

The ants have the ability to find the shortest path going from nest to a food source without use of a visual cue. In addition, there are features to adapt to changes around them. For example, they find again the shortest path in case of the occurrence of any problems on the path going to a food (such as the emergence of a barrier) or if the road is not available. Figure 3.1 shows the ants going to food throughout a linear way.

M. Ünal et al.: Optimization of PID Controllers Using AC and GA, SCI 449, pp. 31–35.
springerlink.com   © Springer-Verlag Berlin Heidelberg 2013

Fig. 3.1 The path of the ants

The means is pheromone to find this way of the ants. Ants help the next ants to choose the path by dropping off stored Pheromones on the chosen way. This instinctive behavior explains how they had found the shortest path to their food, even in the case pre-existing path cannot be used. In fact, if obstacle is on the path to food, the ant in front of this obstacle cannot continue and has to make a choice for a new way (Figure 3.2).

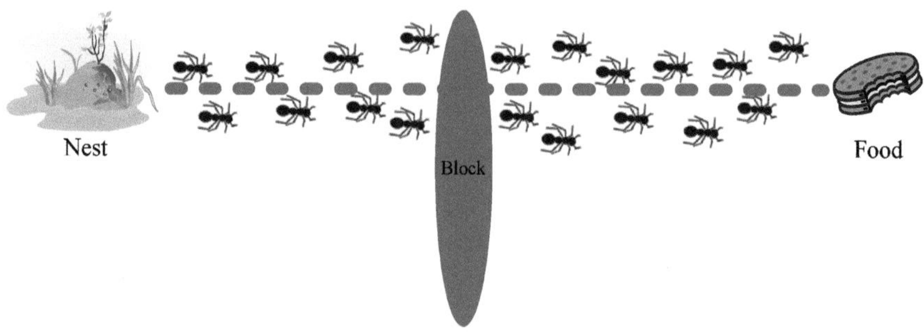

Fig. 3.2 Encountering Ants to an obstacle

In this case, designation probability of the ant's new pats is equal. Namely, if ants can select the right and left direction, selection chances of these directions are equal. Ant goes on its way according to his election and designates his own path as shown in Figure 3.3.

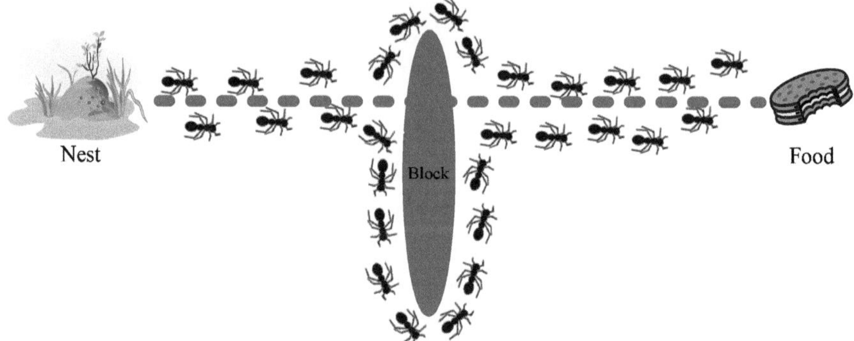

Fig. 3.3 Selection of ants coming across the obstacle

One of the interesting points is if the path chosen by ants isn't the shortest path to the food, ants that choose the road can configure the path again as shown in Figure 3.4.

Fig. 3.4 Find the shortest path by ants

The ant elections will increase the amount of pheromone on the road and would be preferable for subsequent ants. Interaction between and the road must be fast to form the positive effect of pheromone.

Choosing the shortest way after catching the obstacle of the ant may take longer from normal process regarding as each ant left same amount of pheromone and the same speed at average [14].

3.2 Artificial Ants

Real ants, although they are blind can find the shortest path to food slots. This feature of ants has been made available to solve real problems just by using a number of features and some additions. Features from real ants creating artificial ants:

- Communication between the ants established through pheromones,
- Primarily preferring the roads with much pheromones priority over,
- Fast increasing the amount of pheromones on the short roads.

Added features to real ants:

- They live the environment where time is calculated discretely
- They are not totally blind and they can access to details about the problem,
- They can hold information for solving problems with a certain amount of memory.

The basic idea of Ant-based algorithms is that artificial intelligent agents using a simple communication mechanism can produce the solutions to many complex problems [51].

3.3 Ant Colony Algorithm

ACO algorithm is very suitable for parallel working. There is more than ant working asynchrony or synchrony. Ants can be described as intelligent agents in parallel systems assuming that there is more than one node in parallel ant colony system and each ant work a node at a specific time. Ants in the travelling salesman problem visit all the cities starting from selected randomly leave the pheromone to the passing roads. These pheromones are effective in identifying the ways of next ants, i.e. communication between the agents, gives a common result. While each ant is sent a node, one thing to consider is prevent the contest conditions. The contest is meant that the ants working at the same time (the ants running on the different nodes) can't change global data structure that holds pheromones and the selected path data at the same time. Because the main purpose of this study is to find the most appropriate values of parametric alpha (α), beta (β), ro (ρ) used in the probability calculation in ACO algorithm, increasing the number of ants in colony will increase the probability of finding the most appropriates of the values [14].

Each ant is placed different or the same corners at the beginning of the problem. Equation (3.1) (probability equation) determines these ants will situate on which adjacent node at time (t).

$$P_{ij}^{k}(t) \left\{ \begin{array}{ll} \dfrac{\left[\tau_{ij}(t)\right]^{\alpha}\left[\eta_{ij}(t)\right]^{\beta}}{\sum_{i\in N_i}\left[\tau_{ij}(t)\right]^{\alpha}\left[\eta_{ij}(t)\right]^{\beta}} & \left(\begin{array}{l} if\ the\ k\ is \\ an\ allowed \\ selection \end{array}\right) \\ 0 & (otherwise) \end{array} \right\} \qquad (3.1)$$

$\tau_{ij}(t)$: At time t trace amount of pheromones at (i,j) the corners

η_{ij} : Visibility value between (i, j) corners. This value varies according to the criteria taken into consideration in solving the problem

α : Shows the relative importance of the pheromone trace in the problem

β : Shows the importance that given to visibility value

N_i : Set of the node points that hasn't chosen yet

Ants make the next elections according to this possibility equation. A round or iteration is completed after all the nodes at the problem are visited. At this point, amount of pheromone trace is updated according to the Equation (3.2) [50].

$$\tau_{ij}(t+n) = (1-\rho)\tau_{ij}(t) + \Delta\tau_{ij}(t) \qquad (3.2)$$

p : Proportion of pheromone trace evaporated between t and t+1 time period ($0 < p < 1$)

$\Delta\tau_{ij}$: Shows the amount of pheromone trace of the corner due to the election of the (i, j) corner during a tour of the ant. This amount is computed by Equation (3.3).

$$\Delta \tau_{ij}(t) = \sum_{k=1}^{m} \Delta \tau_{ij}^{k} \qquad (3.3)$$

m : Total ant number

$\Delta \tau_{ij}^{k}$: Amount of pheromone trace left by k. ant at (i, j) corner

Equation (3.4) shows the contribute amount of the (k) ant to the pheromone trace at the any (i, j) corner[55].

$$\Delta \tau_{ij}^{k} = \left\{ \frac{Q}{L_k} \right\} \qquad (3.4)$$

Q : A constant

L_k : Tour length of k. ant.

If the ant used the (i, j) corner along the tour, trace amount is calculated according to Equation (3.4). Otherwise, the trace amount is zero [55].

3.3.1 Pheromone Vaporization

Pheromone density decreases due to evaporation and degradation at real ant colonies. Effect of evaporation at the ACO is simulated with the application of the properly defined evaporation rule. For example, artificial delay of pheromone can be set to a fixed rate. Pheromone evaporation reduces the effect of low - quality pheromones created by artificial ants at the early stages of the research. Pheromone evaporation can be very useful on artificial ant colonies even if it has not a noticeable effect on real ants.

Chapter 4
An Application for Process System Control

The purpose method is implemented to stabilize the pressure of the tank at the desired pressure level adjusting the input air flow despite the continuous exhaust output flowing as a disturbance. Because of compressibility of air and nonlinear characteristic of valves, realized system has nonlinear dynamics. Cubic trajectory function was used as an input reference signal, to prevent the pressure fluctuations and large overshoot in tank which could be harmful in some process [56].

This cubic trajectory is defined as follows (Figure 4.1);

$$
\begin{aligned}
p_Y &= p_0 & t &= t_0 \\
p_Y &= \varphi_3 t^3 + \varphi_2 t^2 + \varphi_1 t + \varphi_0 & t_0 &< t < t_s \\
p_Y &= p_s & t &\geq t_s
\end{aligned}
\tag{4.1}
$$

Where p_0, p_s are initial and desired pressure values, $\varphi_{0\text{-}3}$ are trajectory coefficients, t_o is initial time, t_s is setting time defined as $3\alpha \dfrac{P_y - P_0}{P_y + P_0}$ and α is trajectory slope.

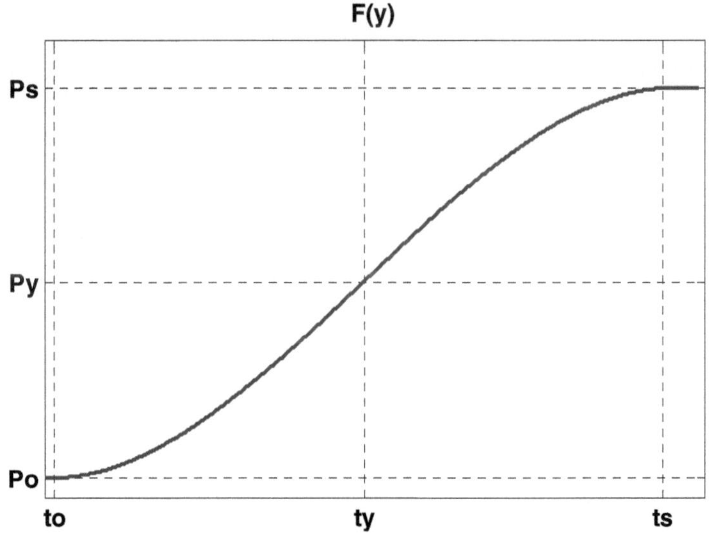

Fig. 4.1 Cubic trajectory

M. Ünal et al.: Optimization of PID Controllers Using AC and GA, SCI 449, pp. 37–68.
springerlink.com © Springer-Verlag Berlin Heidelberg 2013

Realized system and its block diagram are given in Figure 4.2 and Figure 4.3. The tank input and output flows are controlled by pneumatic and manual valve respectively. The output of DAC (4-20 mA) is converted to pneumatic signal via I/P converter to feed the pneumatic valve (Samson 241-7).Tank pressure is measured by P/I (Endress-Hauser) converter and it is used as a feedback signal to the ADC. The MATLAB based PID digital controller is used to control the pressure in the tank. The computer with Advantech PCI 1711 model data acquisition board (DAQ) [57] is used in the hardware platform of the digital controller.

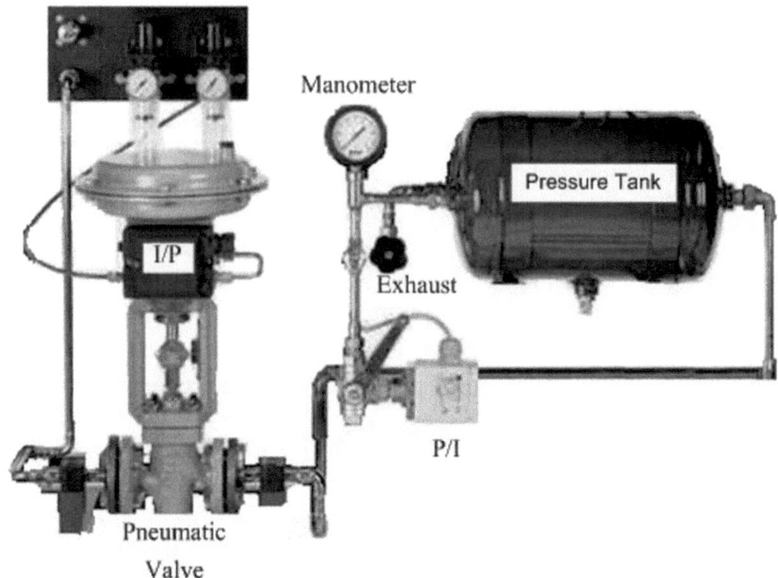

Fig. 4.2 Realized System

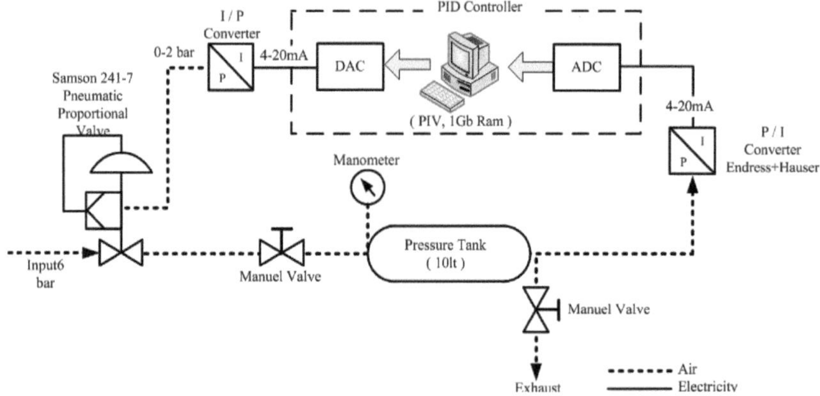

Fig. 4.3 Process Block Diagram

4.1 Modeling of the System with ANN

ANN is used to obtain the dynamic model of the pressure tank composed air intake and discharge from the proportional valve. Serial-parallel ANN model based on time series modeling (NARX) is selected to model behavior of the air pressure control.

4 time delays are used for the application and previous 4 inputs and previous 4 outputs of the ANN were used as inputs of the network.

Designed ANN architecture has two inputs for process output (y), input (u) and one output for predicted value $\hat{y}$. Hyperbolic tangent and linear activation function were used in hidden and output layer. Designed NARX type ANN architecture was given in Figure 4.4.

Once designed ANN was trained, the network can be used for modeling the output for any input vector from the input space. This is called the "generalization property" of the network. To show this property of trained networks testing data set was used. We also used the early stopping strategy to improve the generalization properties of designed network.

To design better ANN applications, the number of hidden layer and neurons in the hidden layer(s) play very important roles and the choice of these numbers depends on the application. Determining the optimal values of these numbers is still a question to deal with. Although there is no theoretical basis for selecting

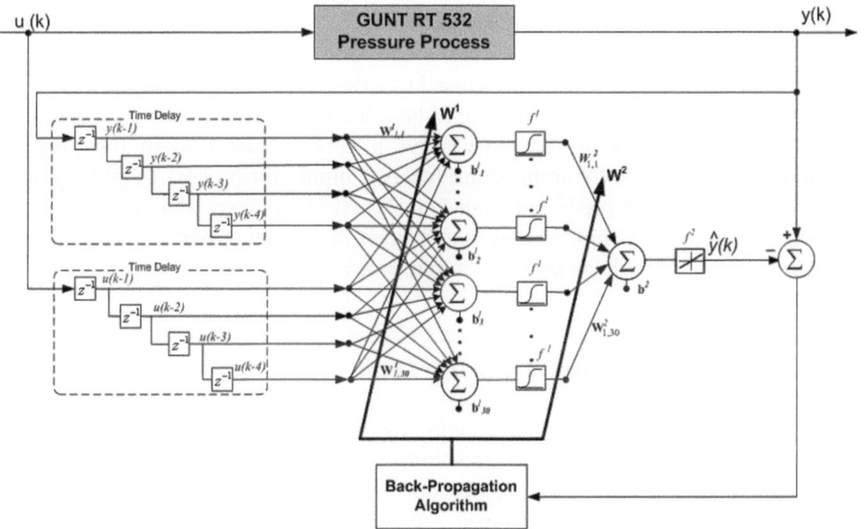

Fig. 4.4 Designed NARX type ANN Structure for Modeling the Dynamic Model of Pressure Process System

Table 4.1 MSE and Correlation Coefficient (R) Values at Different ANN Models

Model	Hidden Layer Neurons Number	MSE Generalization	Correlation Coefficient (R)	
			Learning	Generalization
ANN-5	5	9,90E-06	0,975	0,778
ANN-10	10	8,20E-06	0,983	0,855
ANN-20	20	9,93E-07	0,987	0,892
ANN-30	30	3,70E-07	0,999	0,997
ANN-50	50	3,60E-07	0,999	0,998

these parameters, a few systematic approaches are also reported. But the most common way of determining this number is still trial and error approach. End of the modeling study, MSE and correlation coefficient (R) were given at Table 4.1. Higher values of correlation coefficients and lower MSE values show that model and its predicted values are closed together. According the table 1, ANN-30 network has satisfactory performance in both learning and generalization phase.

Once designed ANN was trained, the network can be used for modeling the output for any input vector from the input space. This is called the "generalization property" of the network. To show this property of trained networks testing data set was used. We also used the early stopping strategy to improve the generalization properties of designed network.

Higher values of correlation coefficients and lower MSE values show that model and its predicted values are closed together. According the table 1, ANN-30 network has satisfactory performance in both learning and generalization phase.

The NARX model of the plant plays a very crucial role in the proposed structure. Designed ANN architecture has two inputs for process output (y), input (u) and one output for predicted value ($\hat{y}$). Hyperbolic tangent and linear activation function were used in hidden and output layer. Time delay unit described as (z^{-1}) block and input time delay (m) and output time delay (n) were selected as four and five respectively.

4.1.1 Collection of the Training Data

It is needed to collect training data including different working areas of the system for generated model close to the real system. Training data are pressure information obtained by applying random inputs and 0-10V step inputs to the system. Applied inputs and obtained pressure data are shown in Figure 4.5 and Figure 4.6.

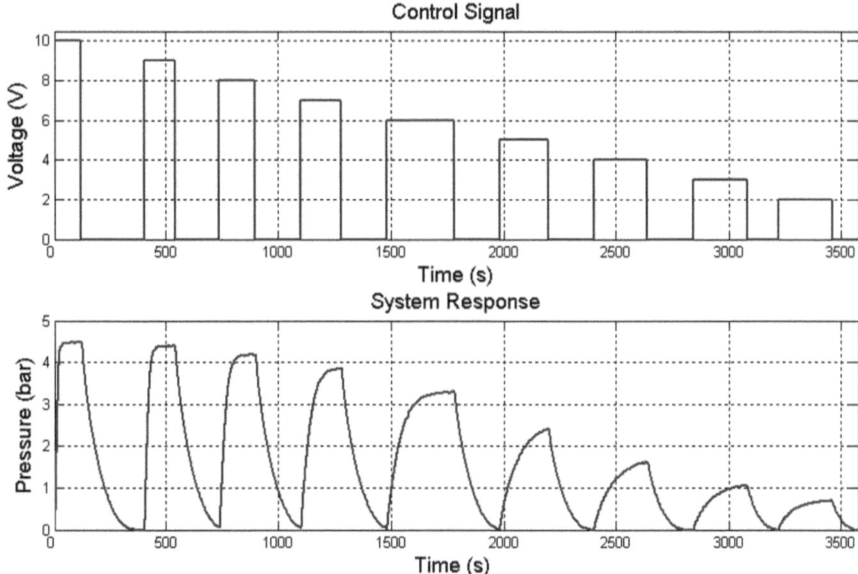

Fig. 4.5 Step inputs applied for NN training and sytem response

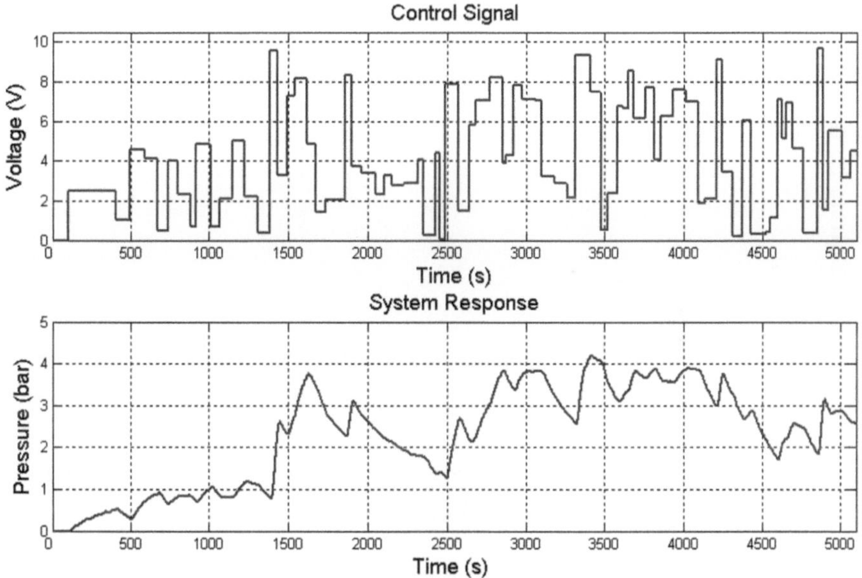

Fig. 4.6 Random inputs applied for ANN training and sytem response

4.1.2 Data Normalization

Normalization process is a scaling to improve the accuracy of result in numerical calculation of the numbers composing data set. Normalization process is the important stage for the ANN training. In addition, it adapts the activation function working range. Equation 4.1 can be used for normalization process.

$$y = \frac{x - x_{min}}{x_{max} - x_{min}}$$

(4.2)

where y is the normalized output, x is the output value, x_{min} is the minimum output value, x_{max} is the maximum output value.

4.1.3 NN Training

Levenberg-Marquardt training algorithm is used to train generated ANN model. Training parameters were selected as the number of iterations 50 and MSE error 10^{-6}. ANN model training was terminated at the end of 41st iteration because MSE error falls below 10^{-6}. Change of MSE through training steps shown in Figure 4.7.

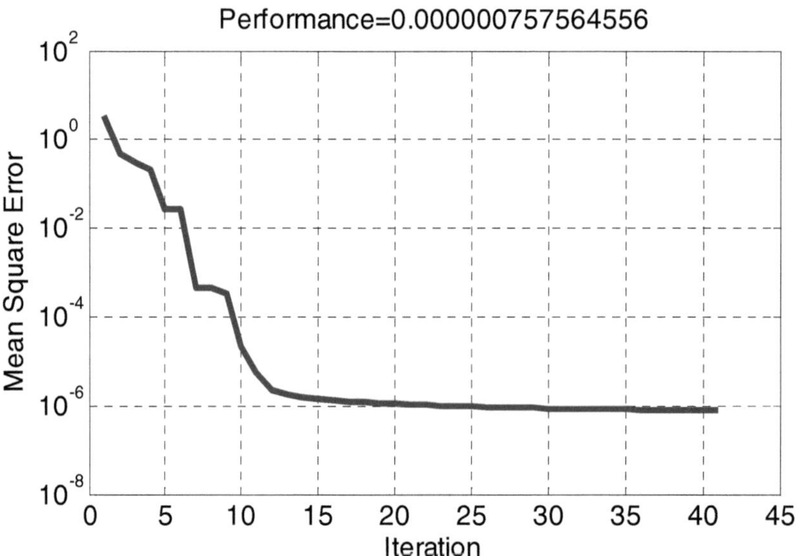

Fig. 4.7 Change of MSE through training steps

Regression curve of the system output and the trained ANN model output is shown in Figure 4.8. ANN model output is successfully converged as the desired target output at the end of training as shown in Figure 4.8.

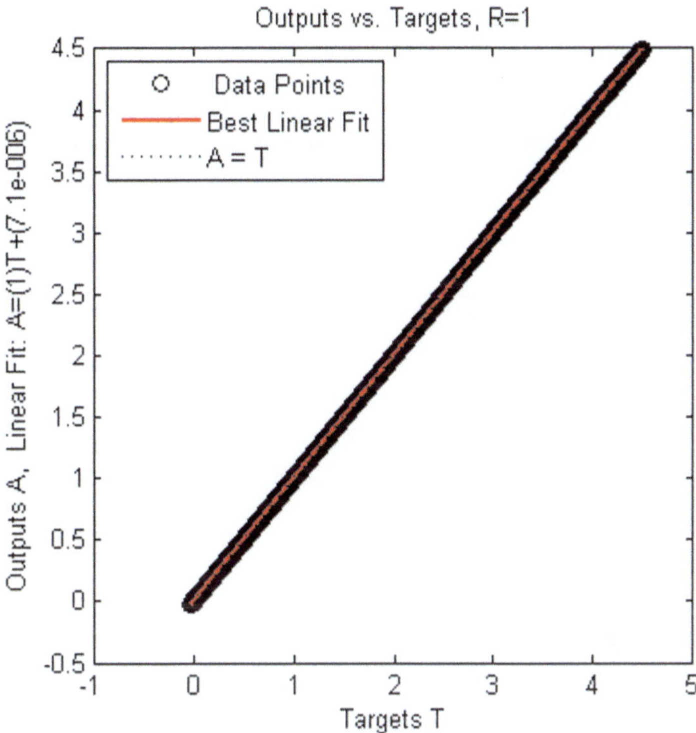

Fig. 4.8 Regression curve of the target data and the ANN model output

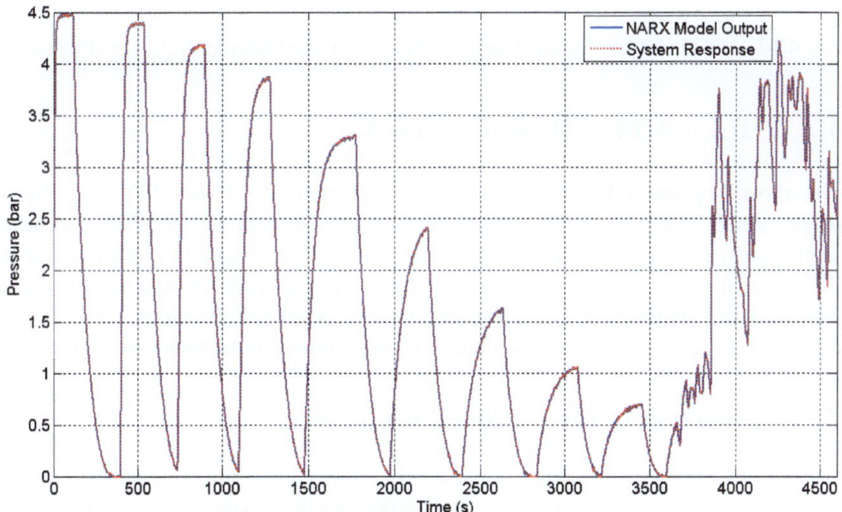

Fig. 4.9 System and ANN Model output graph

Pressure change obtained applying 0-10V control signal to the system and ANN model output graphs are shown Figure 4.9.

So, show the ANN performance clearly, system open-loop outputs and ANN outputs at different pressure values were given in Figure 4.10. As could be inferred from this figure, ANN model and real system behavior are consistent and quite similar.

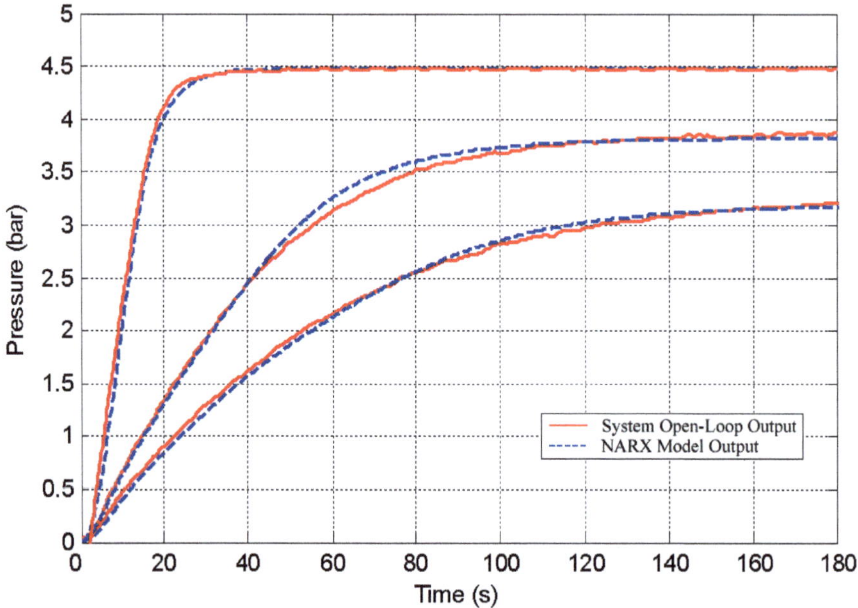

Fig. 4.10 ANN model and System Open-loop Output at the Different Reference Pressure

4.2 PID Controller Design with GA

The most important factor affecting the performance of the PID controller is proportional, integral and derivative action coefficients of the PID controller. There are mathematical and experimental methods for finding coefficients of the PID controller. Mathematical model of the system must be present for the use of mathematical methods. Controlling the system within the desired limits is very difficult on experimental methods and it's needed to be done many experiments.

4.2.1 Structure of the Designed Controller

General structure of the system conducted for pressure control unit off-line study is shown in Figure 4.11. Add-ons were made to the system control for the safe

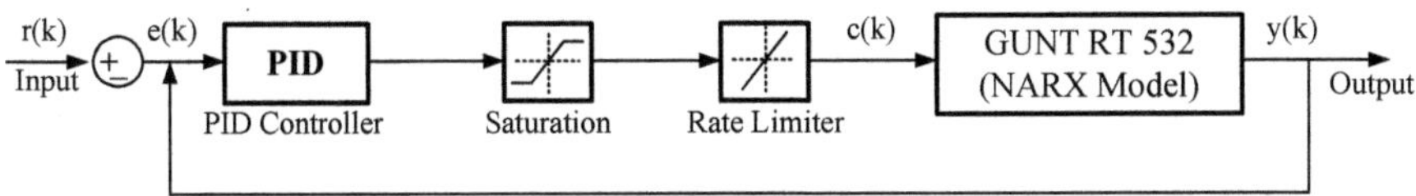

Fig. 4.11 General closed structure of the used for off-line study

operation of the system. Output of the PID controller is limited to 10V and valve movement is slowed to not show a very rapid change of the pneumatic valve on controlling the system by using rate-limiting.

Controller produces the error signal single input and single output (SISO) system with negative feedback in Figure 4.11 shown in block diagram, comprising a system reference to the by taking the difference of the output of from the control signal according to the manufactures.

Block diagram of the genetic-PID controller designed for this search process is shown in Figure 4.12.

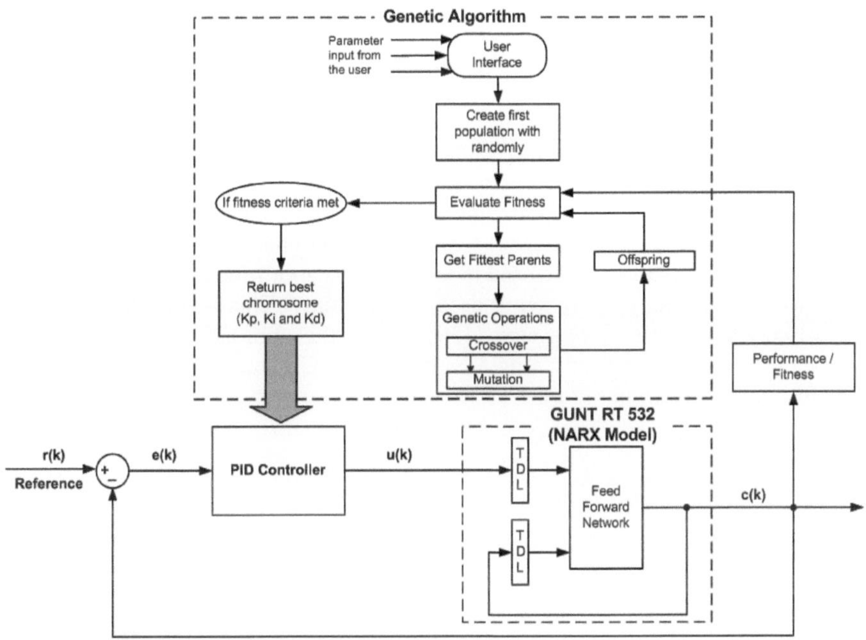

Fig. 4.12 Optimization of the PID controller with GA

To perform such a control system, three major problems must be resolved. These are:

- Genetically coding of the PID parameters,

- PID parameters learning method of genetic algorithm,
- Converting the output performance of the system into genetic algorithm compliance value [58].

4.2.2 Genetically Coding of the PID Parameters

Determination of the minimum values of the PID controller Coefficients and limitations of the PID controller output is required before PID controller coefficient found by GA in offline operation mode. The lower limit value is set to zero coefficients of the PID controller on designed system.

GA operates on the encoded form of the problem not by itself. Therefore, coding format of the problem that will be solved has a very significant impact on the performance of GA. There are two types coding in the literature most commonly used: Weighted Binary Encoding and the real coding. In this study, values the PID coefficients are encoded with real number as form a single chromosome [59].

In this form of encoding, three parameters are sufficient: proportional (K_p), integral (K_i) and derivative (K_d) for the PID controller. Here, considering each parameter as a gene, 3 genes are needed.

4.2.3 Learning of the PID Parameters by GA

GA parameters are called control parameters. Control parameters can be considered as population size, crossover probability, mutation probability, generation gap, the selection strategy and the function scaling [60].

GA parameters, has a significant influence on the performance of genetic algorithm. There are a lot of works to find the optimal control parameters. But general parameters that can be used in all problems aren't found [61].

Pittsburgh learning approach is used in this study. This approach works very effectively in Genetic fuzzy systems [59]. To use this approach, the PID controller coefficients are encoded in a single chromosome. Chromosome structure created by this way is shown in Figure 4.13.

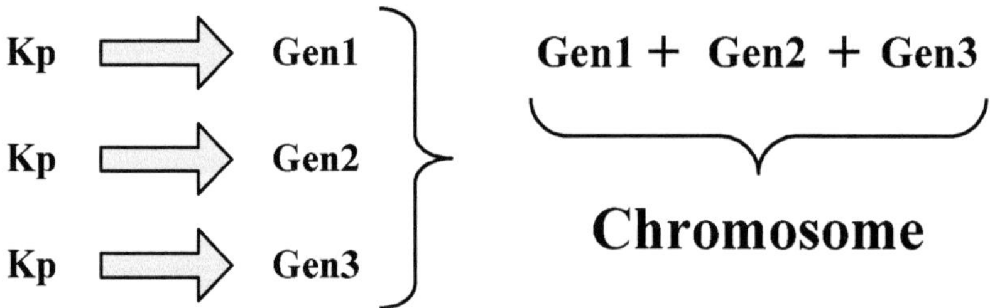

Fig. 4.13 Chromosome structure created for PID Controller

4.2.4 *Transformation of the System Output to GA Conformity Value*

In this study, because it's required to trace the system output along a trajectory, a fitness function is created to minimize the error that will be comprise along this trajectory. This adaptation function is as following.

$$\phi = \frac{\sum_{k=1}^{n} \sqrt{\left(r_k - y_k\right)^2}}{n} \tag{4.3}$$

where Φ is adaptation value, k the order of the data depending on sampling time and operation on time at off-line study, n is the total number of data, r is references data and y is the data of the system output. In this equation, controller's performance has been transformed to adaptation value for GA by using the RMS error value.

The PID coefficient where RMS error between the system output and the reference is the minimum according to the adaptation values obtained from Equation (4.3) is obtained by GA. Thus, PID Controller system is followed by the reference with minimum error.

4.2.5 *Used Genetic Operators*

Basic genetic operators used in GA are the coding, population size, crossover, mutation and selection. Genetic operators and the values used in the system realized are:

Coding: In this study, real number coding was chosen and values of the PID coefficients are encoded with real numbers to form a single chromosome.

Population size: Population size is quite influenced by GA's performance. Use of processor's values and finding the result time is quite increased for high population size. GA cannot find the desired result for very small values of the population. So different experiments were made and the most appropriate population size is chosen as $N = 20$.

Crossover: Arithmetic crossover operator is used. Probability value of crossover operator is selected as 0.8.

Mutation: Gaussian mutation operator is used. The mutation operator scale and rate of contraction is chosen as 1.

Reproduction: Shifting reproduction operator is used for GA structure in this study. In addition to this method elistic model is used. Elistic model is the transfer of four populations with best probability value to the next generation without any change. In this way, death of the most appropriate chromosomes is prevented without being transferred to the next generations. The performance of the GA is increased with Elistic model.

Election: Tournament method was used for the selection operator. The tournament size was selected as 4 in the experiments.

System was made oscillation with procreating of the incompatible PID coefficients, i.e. consists of bad individuals and steady state error was occurred.

4.2.6 Performance of the Genetic-PID Controller

Real time performances of the genetic-PID controller obtained during optimization with ANN model are compared in this section. When PID coefficients in the ANN model used in real-time system control, successful results were obtained as can be seen in the results.

4.2.6.1 System Response to Step Inputs

Reference of 1 bar as shown in Figure 4.14 and 3 bar as shown in Figure 4.15 are given to system, then ANN model and actual system outputs following these reference values are shown. Control signals providing these outputs are shown at the top of the figures. It can be understood from the behavior of the control signals, when system output approaches the reference value, PID controller was abandoned driving the system on maximum and started to produce control signal for steady-state.

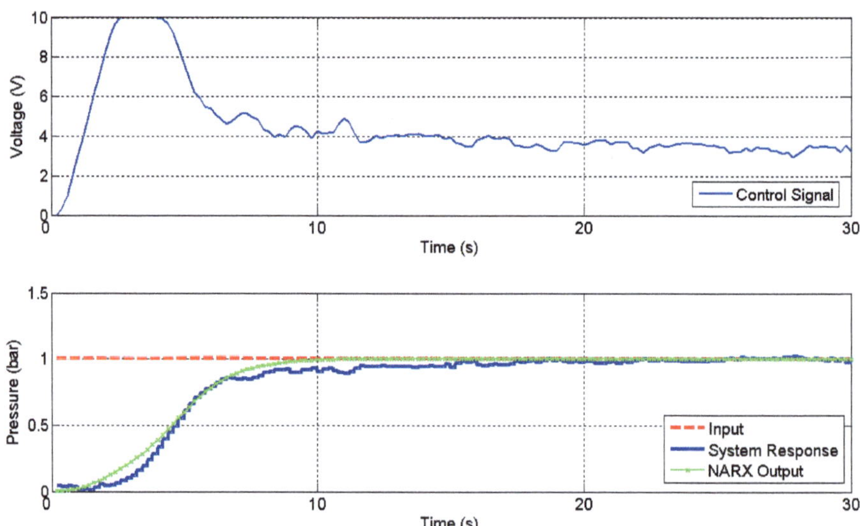

Fig. 4.14 Outputs of the Control, ANN Model and Actual System for reference input of 1 bar

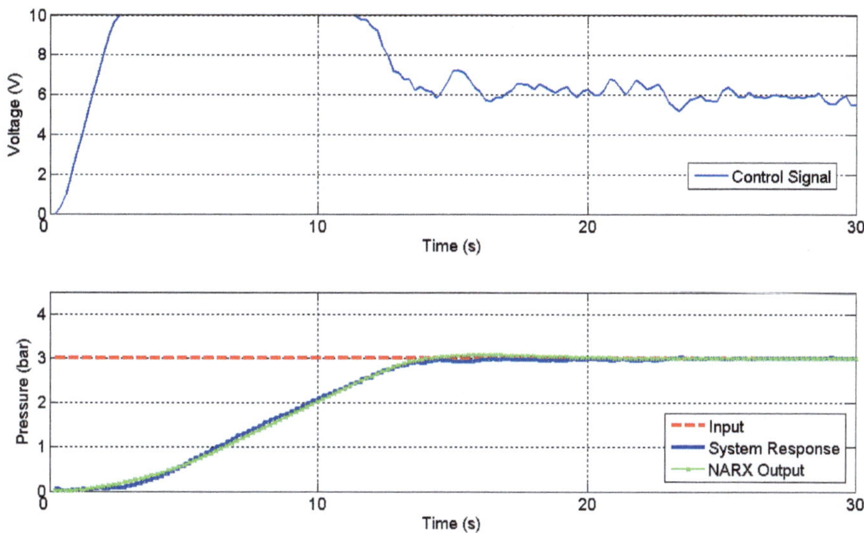

Fig. 4.15 Outputs of the Control, ANN Model and Actual System for reference input of 3 bar

4.2.6.2   System Behavior to the Transition between Step Inputs

In this section, to see the transition response of the genetic-PID controller to the different references, system response is investigated giving different references. After system located the first reference value, transition of the system between the references is investigated. Graphics of this study are given at Figure 4.16. System

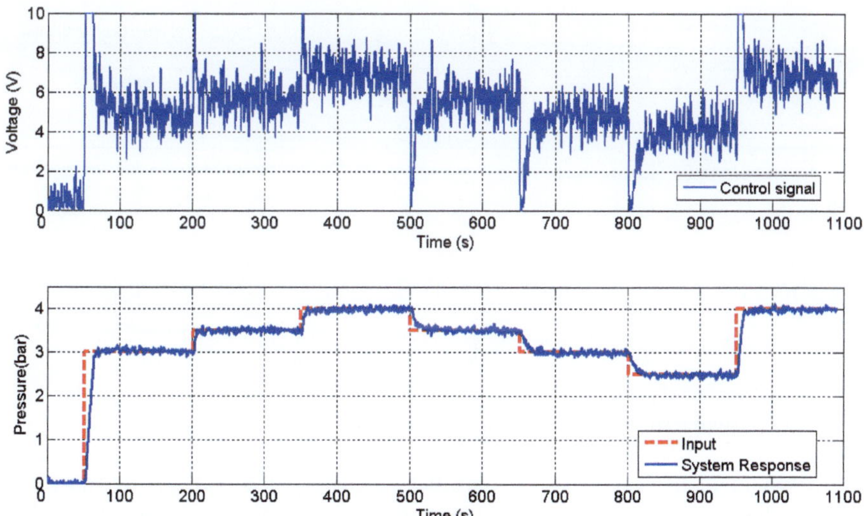

Fig. 4.16 Actual system response during step transitions

successfully carried out the transition between different inputs as shown in figure. So, system has been successfully controlled with PID coefficients obtained with GA on real-time.

4.2.6.3 Behavior of the System When Distortion Was Imported to the System

In this part of the study performance of the purposed genetic-PID controller is carried out, in the case of a disturbance imported from outside.

During the operation of the system, with applying 6 V voltage to the proportional pressure valve at the pneumatic tank output, opening of the valve in the rate of 50% is provided by. Thus, evacuation of the much air is provided while the system is working. Disturbance has been imported with increasing the space value at the output. Discharge rate of the air in the tank increases by giving voltage more than 6 V. Thus, a negative disturbance has been imported to the system. Behavior of the genetic-PID controller is investigated with giving more than 25% voltage to the proportional valve at the system output on Figure 4.17. Here, after the system set reference of 4 bar, disturbance input was applied for 90 s and then normally working of the system was started. System output was dropped below 4 bar when disturbance input was applied. However, system output reached again 4 bar with raising the control signal from 7.5 V level to 9 V. System output come up 4 bar after disturbance getting out and system output was again placed on reference value of 4 bar with the control signal coming to 7.5 V level.

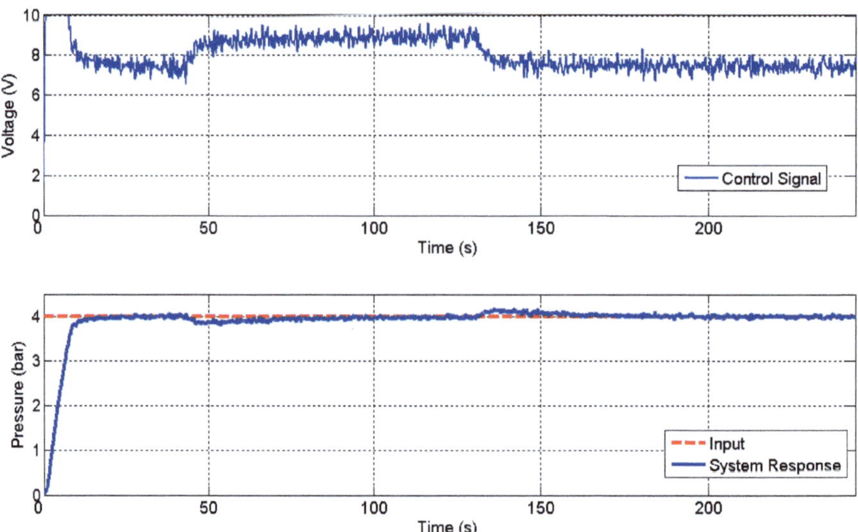

Fig. 4.17 System responses with applying %25 distortion

Behavior of the genetic-PID controller is investigated with giving more than 40% voltage to the proportional valve at the system output on Figure 4.18. Here, after the system set reference of 4 bar, disturbance input was applied for 90 s and then normally working of the system was started. System output was dropped

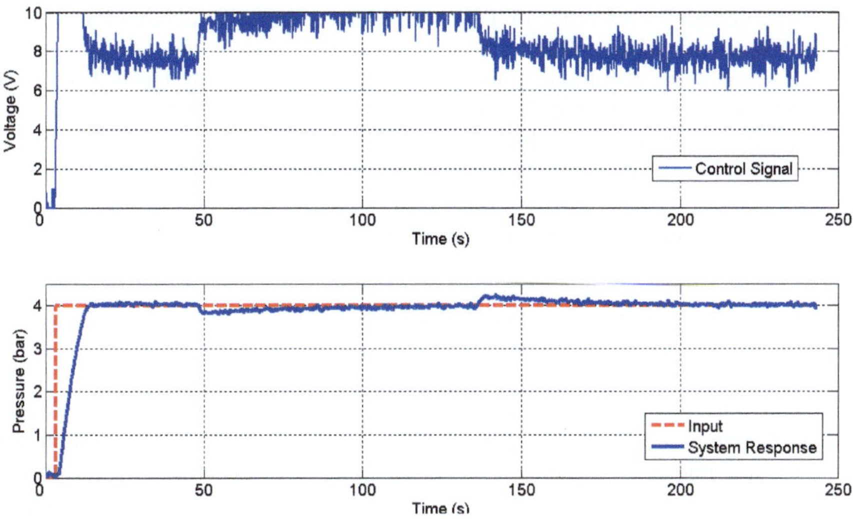

Fig. 4.18 System responses with applying %40 distortion

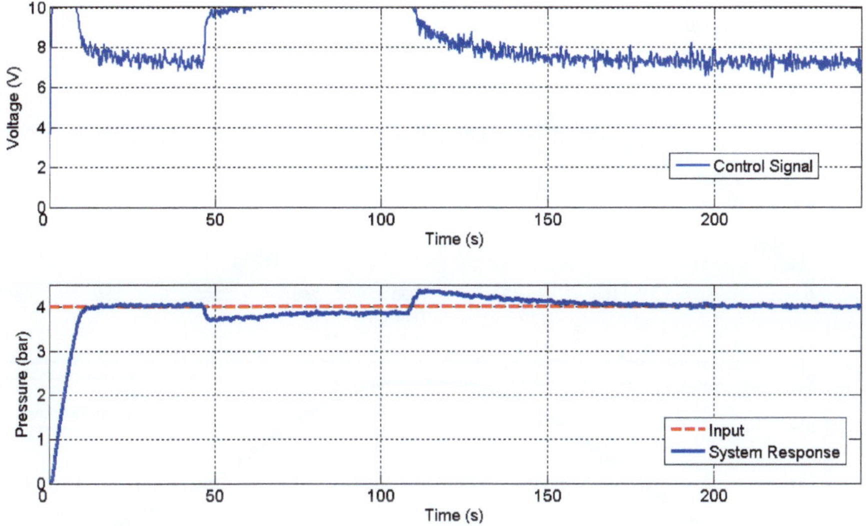

Fig. 4.19 System responses with applying %50 distortion

below 4 bar when disturbance input was applied. However, system output was reached again 4 bar with raising the control signal from 7.5 V level to 9.5 V. System output came up 4 bar after disturbance getting out and system output was again placed on reference value of 4 bar with the control signal coming to 7.5 V level.

Behavior of the genetic-PID controller is investigated with giving more than 50% voltage to the proportional valve at the system output on Figure 4.19. Here, after the system set reference of 4 bar, disturbance input was applied for 70 s and then normally working of the system was started. System output was dropped below 4 bar when disturbance input was applied. However, system wasn't set to reference value even if control signal was 10 V i.e. 100%. System output come up 4 bar after disturbance getting out and system output was again placed on reference value of 4 bar with the control signal coming to 7.5 V level.

4.2.6.4 Trajectory Tracking Responses of the Genetic-PID

In this part of the study, the trajectories created with different grades from 0 bar to 4 bar (with the trajectory function defined in Equation 4.1 given to the system as reference trajectory the system behavior is investigated.

First, the system is purposed to follow from initial value to 4 bar. For this purpose, trajectory is generated with given 6 to the slope and given to the system as a reference. When system came to the 4 bar, response of the system was investigated during the discharge by creating the necessary trajectory for coming of system to the first state with same slope. System response is shown at Figure 4.20. The system was successfully tracked the trajectory during charging. During the last sections, a difference was occurred between the trajectory and system response, because natural discharge of the system is not sufficient to track the trajectory although control signal is 0 after 300 s.

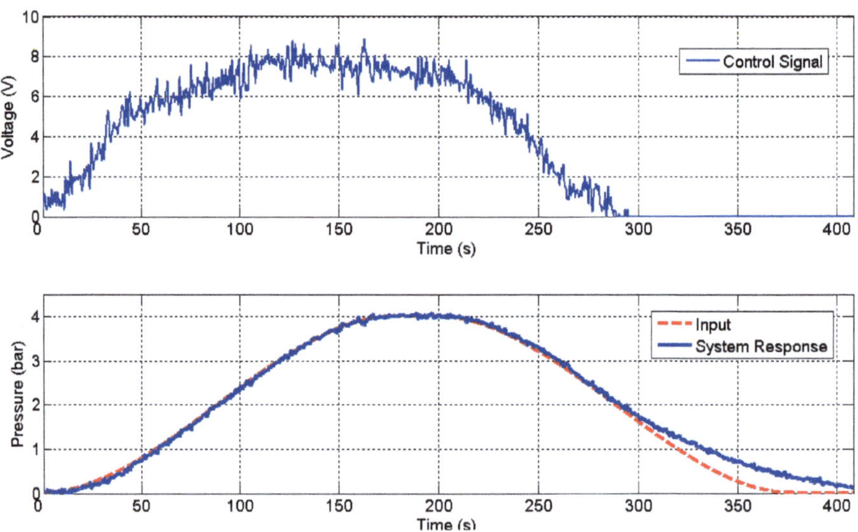

Fig. 4.20 Tracking of the reference trajectory with slope 6 by the system

Charging and discharging trajectories are combined by the slope 15 then given to system as reference. System response to this reference trajectory is given in Figure 4.21. It is understood from this graph that the system tracks the trajectory during charging and discharging successfully.

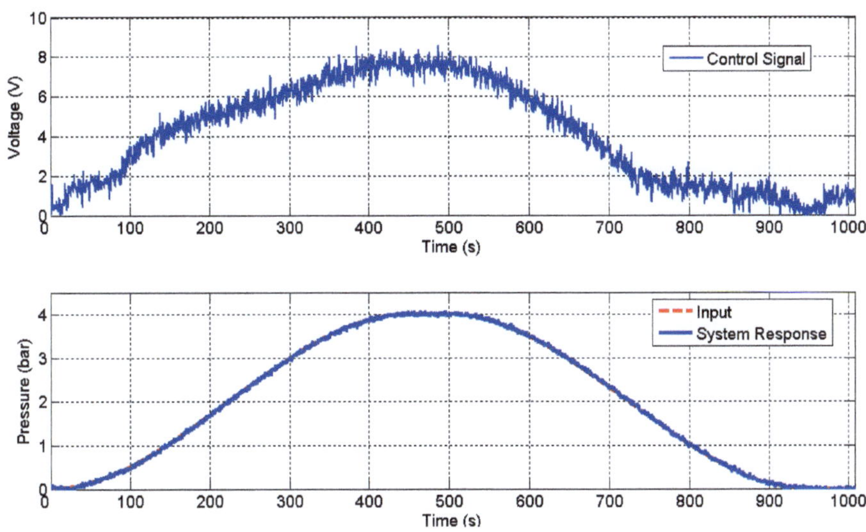

Fig. 4.21 Tracking of the reference trajectory with slope 15 by the system

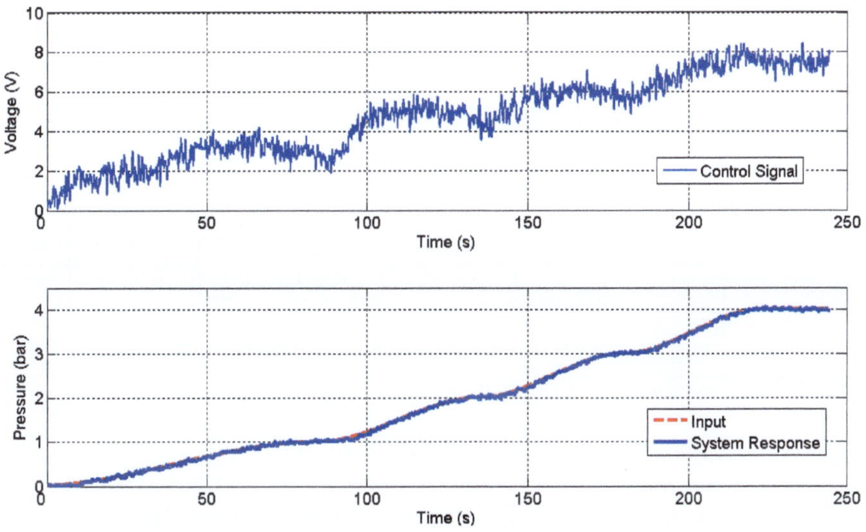

Fig. 4.22 System response to transition between trajectories

As can be understood from the graph, the system tracks successfully the trajectory during charging and discharging.

A reference trajectory as shown in Figure 4.22 was given to see success of the genetic-PID controller during the transition of system between trajectories with different slopes. System successfully carried out the transition between trajectories with different slopes.

As is evident from the obtained graphics, in off-line study, real-time system control with the PID controller optimized by GA carried out successfully even if different references is applied.

4.3 PID Controller Design with ACO

Many researchers have been working on PID tuning methods with ACO up to now and they used different kinds of approach to improve the performance of ACO. Varol et al [62] used the ACO to tune PID controller parameter for a second order system with different cost function. They found very satisfactory results over the classical tuning method. Bin et al [13, 15, 50-52] also used improved ACO for nonlinear PID controller and they concluded that their improved approach was very effective. To optimization PID controller with ACO block schema is given in Figure 4.23.

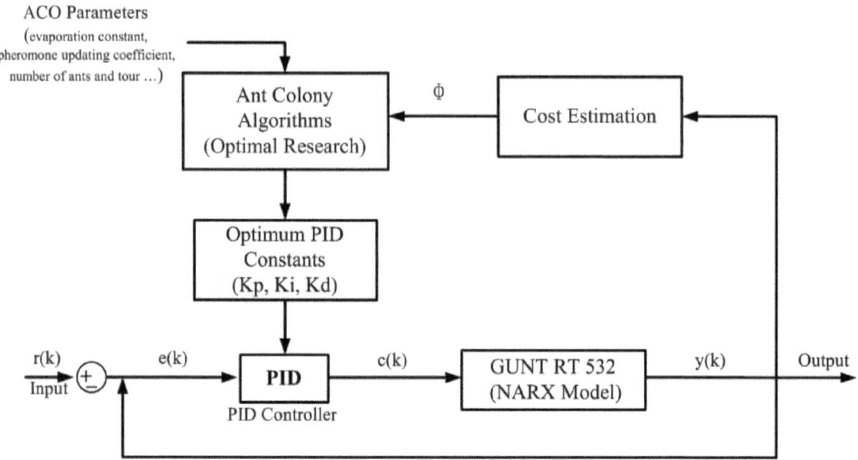

Fig. 4.23 Optimization of the PID Controller of ACO

To solve PID controller design problem with ACO, this problem could be represented as a graph problem which is given in Figure 4.23. All of the values for each parameter (K_p, K_i, K_d) are placed in three different vectors. In order to create a graph representation of the problem, these vectors can be considered as paths between the nests. In the tour, the ant must visit three nests by choosing path between start and end node. The objective of ACO is to find the best tour with the lowest cost function (given in Equation 4.3) among the three nests. The ants deposit pheromone to the beginning of each path. Then the pheromones were updated in pheromone updating rule.

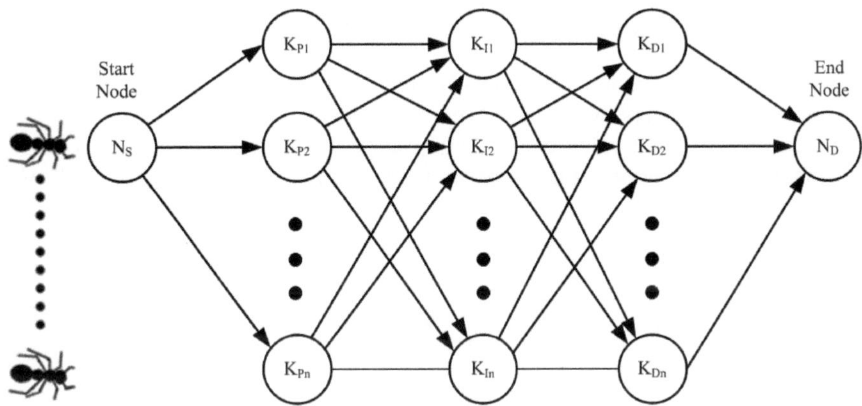

Fig. 4.24 Graphical Representation of ACO for PID Tuning Process

In proposed approach, each ant updates the pheromones deposited to the paths it followed after completing one tour defined as local pheromone updating rules as follows;

$$\tau(k)_{ij} = \tau(k-1)_{ij} + \frac{0.01\,\theta}{J} \tag{4.4}$$

where $\tau(k)_{ij}$ is pheromone value between nest (i) and (j) at the k. iteration, θ is the general pheromone updating coefficient, J is the cost function for the tour travelled by the ant.

In global pheromone updating rule, pheromones of the paths belonging to the best tour (4.4) and worst (11) tour of the ant colony are updated as given in the follows;

$$\tau(k)_{ij}^{best} = \tau(k)_{ij}^{best} + \frac{\theta}{J_{best}} \tag{4.5}$$

$$\tau(k)_{ij}^{worst} = \tau(k)_{ij}^{worst} - \frac{0.3\theta}{J_{worst}} \tag{4.6}$$

where τ^{best} and τ^{worst} are the pheromones of the paths followed by the ant in the tour with the lowest cost value (J_{best}) and with the highest cost value (J_{worst}) in one iteration, respectively.

The pheromones of the paths belonging to the best tour of the colony are increased considerably, whereas those of the paths belonging to the worst tour of the iteration are decreased. After then pheromone evaporation (4.6) allows the ant algorithm to forget its past history, so that ACO can direct its search towards new directions without being trapped in some local minima.

$$\tau(k)_{ij} = \tau(k)_{ij}^{\lambda} + [\tau(k)_{ij}^{best} + \tau(k)_{ij}^{worst}] \tag{4.7}$$

where λ is the evaporation constant [15].

4.3.1   ACO Parameters

Step of PID optimization with ACO is given in Figure 4.25. In this section, PID controller parameters were coded by 5000 nodes. In other words, one node represents a solution value of the parameters K_P, K_i and K_d. Thus, the bigger the number of used nodes, the more accuracy trails are updated. Optimal ACO parameters ($\theta=0.06$, $\lambda=0.95$) were obtained end of numerous experiments.

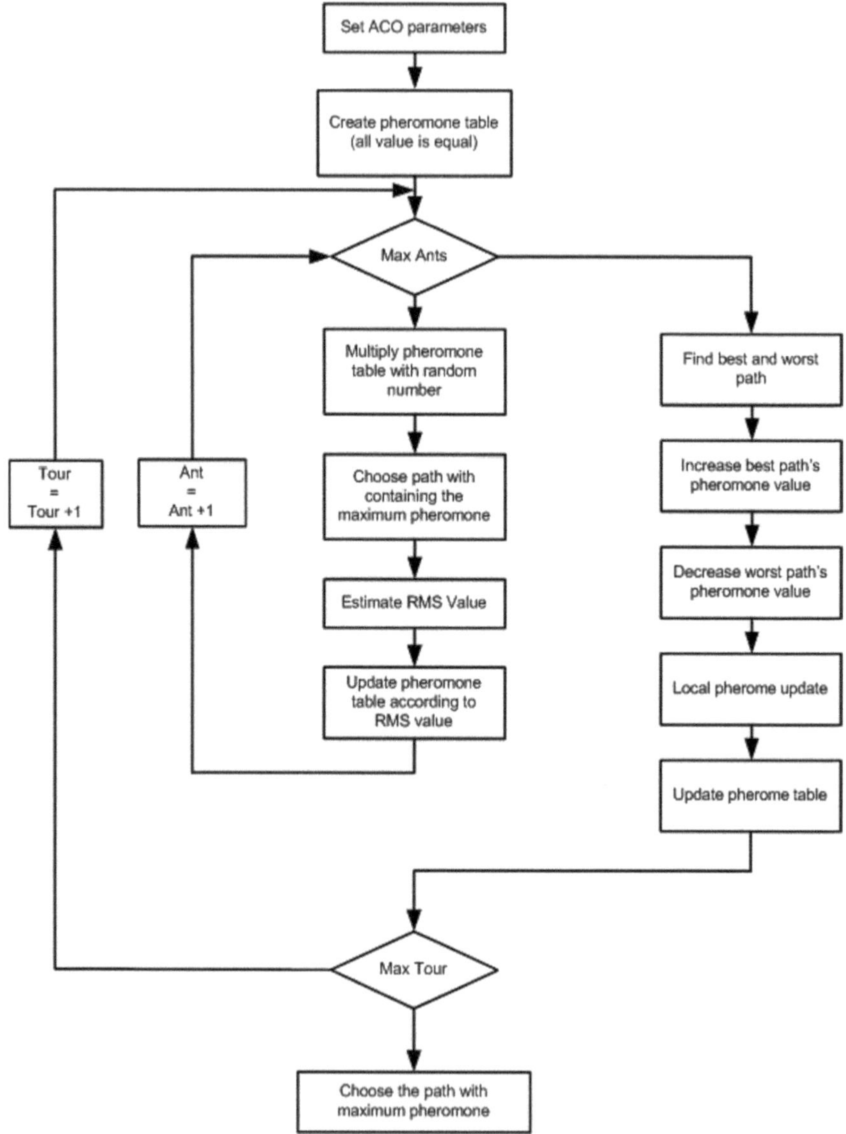

Fig. 4.25 Algorithm for Optimization of the PID Controller of ACO

After ACO parameters were selected, equal number of pheromone were given to the roads that all artificial ants can go and this value was hold at pheromone matrix. When the first artificial ant leaves from the nest, the ant chooses a random path because there is equal amount pheromone at roads that ant have access and complete the tour by visiting three nests. At the end of the tour, simulation was executed with selected PID coefficients and value of roads passing by the artificial ant at pheromone matrix was updated by calculating RMS error between reference pressure input and simulation output. So as to next artificial ants didn't go the same way in the road, pheromone table also was updated by multiplied a random number. Thus, the ants have tried to track the given reference trajectory by different PID coefficients in each tour. When five artificial ants complete the tour, pheromone amount at path of the artificial ant that has least RMS error in this tour is extra increased and pheromone amount at path of the artificial ant that has maximal RMS error in this tour is reduced than amount of the good ant. Also, pheromone amount at the roads that all of artificial ants passed are evaporated with evaporation constant (λ). If the RMS error is resulted small with selected PID coefficients, next ants try to finish their works winding these coefficients. When maximum tour is completed, the roads that have maximum value at the pheromone table, i.e. selected PID coefficients, are recorded as a result of the optimization.

Evolution of ACO cost function on a typical run which was set by optimal parameters was given in Figure 4.26. End of the 100 tour, obtained PID controller parameters were also given in Table 5.1.

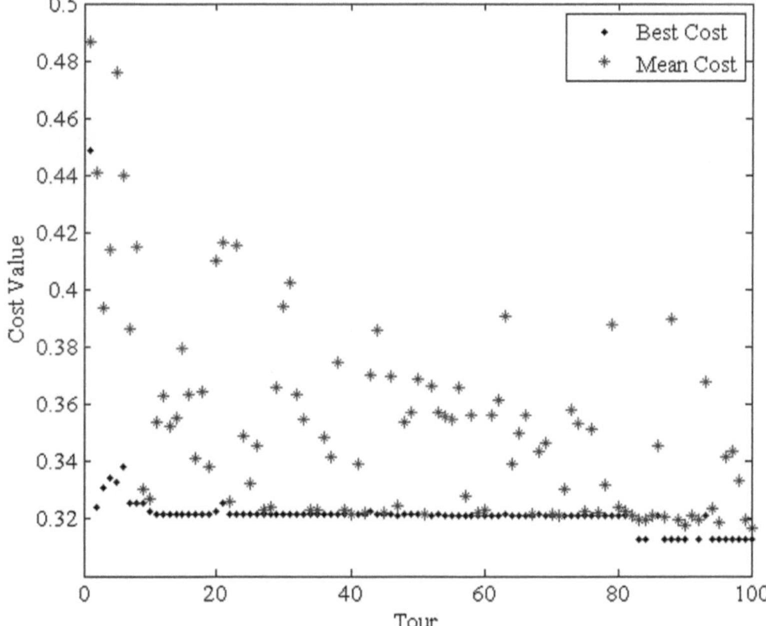

Fig. 4.26 The evolution of ACO cost function

4.3.2 Performance of the ACO-PID Controller

4.3.2.1 System Response to Step Inputs

Reference of 1 bar as shown in Figure 4.27 and 3 bar as shown in Figure 4.28 are given to system, then ANN model and actual system outputs following these reference values are shown. Control signals providing these outputs are shown at the top of the figures. It can be understood from the behavior of the control signals, when system output approaches the reference value, PID controller was abandoned driving the system on maximum and was started to produce control signal for steady-state.

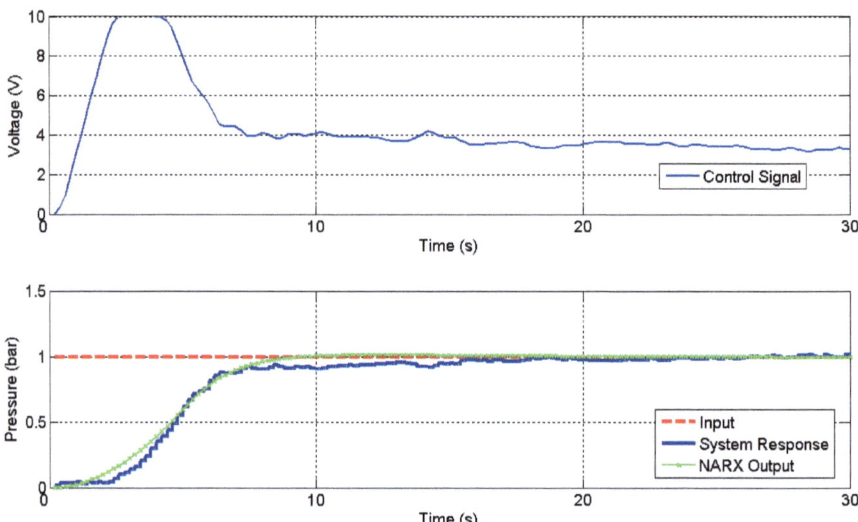

Fig. 4.27 Outputs of the Control, ANN Model and Actual System for reference input of 1 bar

4.3.2.2 System Behavior to the Transition between Step Inputs

In this section, to see the transition response of the ACO-PID controller to the different references, system response is investigated giving different references. After system located the first reference value, transition of the system between the references are investigated. Graphics of this study are given at Figure 4.29. System successfully carried out the transition between different inputs as shown in figure. So, system has been successfully controlled with PID coefficients obtained with ACO on real-time.

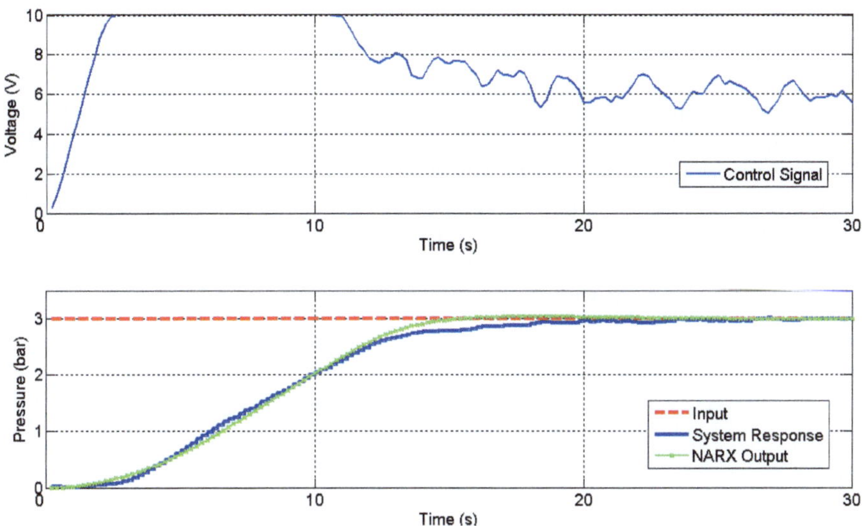

Fig. 4.28 Outputs of the Control, ANN Model and Actual System for reference input of 3 bar

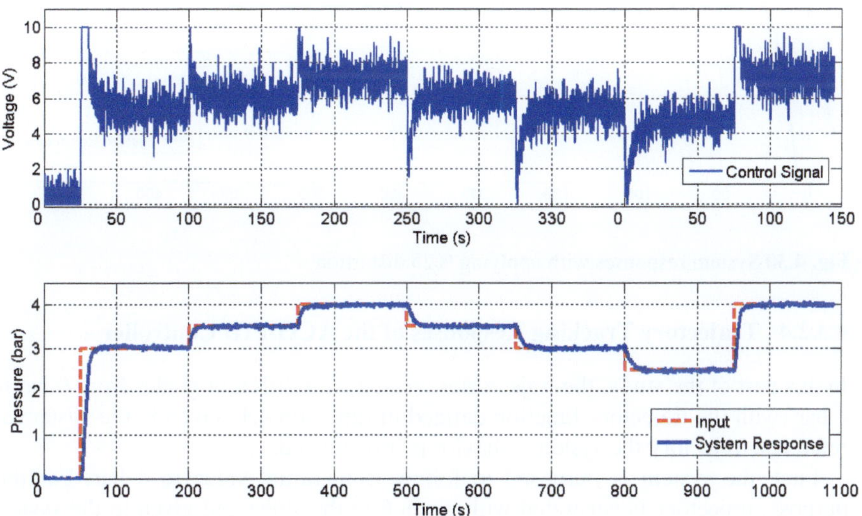

Fig. 4.29 Actual system response during step transitions

4.3.2.3   Behavior of the System When Distortion Was Imported to the System

In this part of the study performance of the purposed ACO-PID controller is carried out, in the case of a disturbance imported from outside. Disturbance has been imported by increasing the opening value of valve at the output. Discharge rate of the air in the tank increases by giving voltage more than 6V. Thus, a negative disturbance has been imported to the system.

Behavior of the ACO-PID controller is investigated with giving more than 25% voltage to the proportional valve at the system output on Figure 4.30. Here, after the system set reference of 4 bar, disturbance input has applied for 150 s and then normally working of the system was started. System output was dropped below 4bar when disturbance input is applied. However, system output was reached again 4 bar with raising the control signal from 7.5 V level to 9 V. System output come up 4bar after disturbance getting out and system output was again placed on reference value of 4 bar with the control signal coming to 7.5 V level.

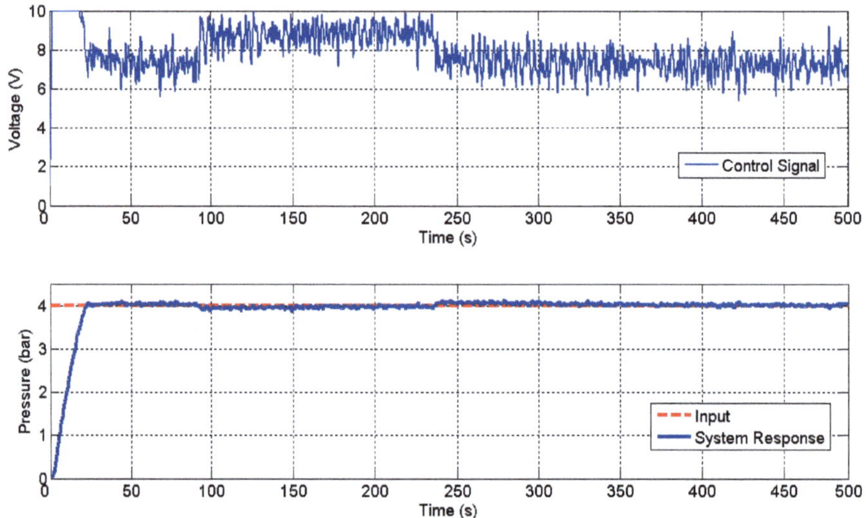

Fig. 4.30 System responses with applying %25 distortion

4.3.2.4 Trajectory Tracking Responses of the ACO-PID Controller

In this part of the study, the trajectories created with different grades from 0 bar to 4 bar (with the trajectory function defined in Equation 4.1) given to the system as reference trajectory the system behavior is investigated.

First, the system is purposed to follow from initial value to 4 bar. For this purpose, trajectory is generated with given 6 to the slope and given to the system as a reference. When system comes to the 4 bar, response of the system was investigated during the discharge by creating the necessary trajectory for coming of system to the first state with same slope. System response is shown at Figure 4.31. The system was successfully tracked the trajectory during charging. During the last sections, a difference was occurred between the trajectory and system response, because natural discharge of the system is not sufficient to track the trajectory although control signal is 0 after 300 s.

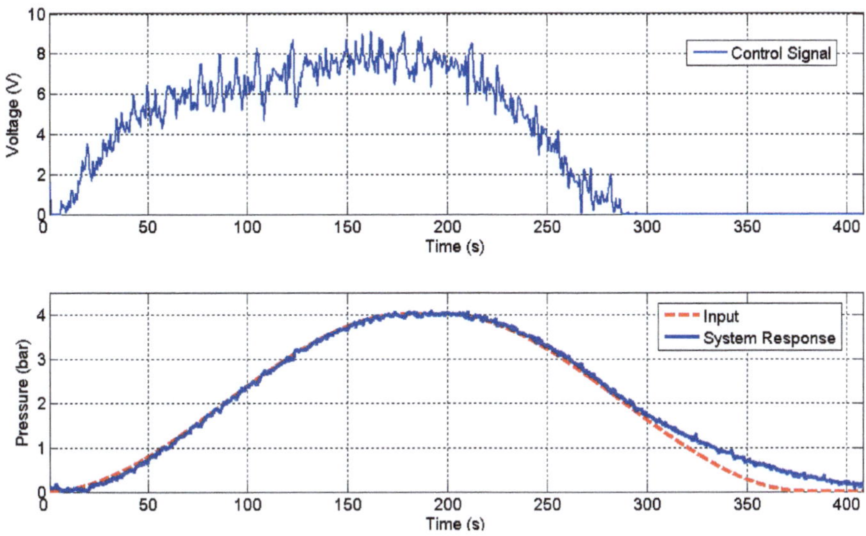

Fig. 4.31 Tracking of the reference trajectory with slope 6 by the system

Charging and discharging trajectories are combined by the slope 15 then given to system as reference. System response to this reference trajectory is given in Figure 4.32. It is understood from this graph that the system tracks the trajectory during charging and discharging successfully.

As can be understood from the graph, the system tracks successfully the trajectory during charging and discharging.

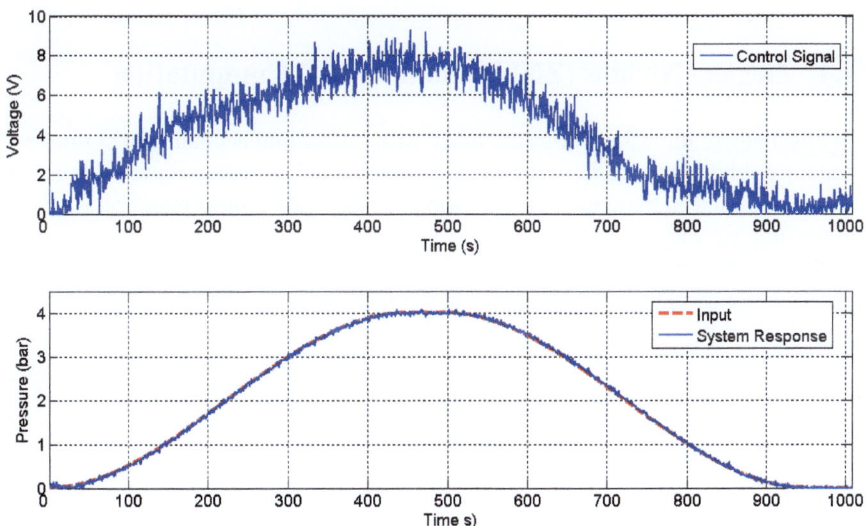

Fig. 4.32 Tracking of the reference trajectory with slope 15 by the system

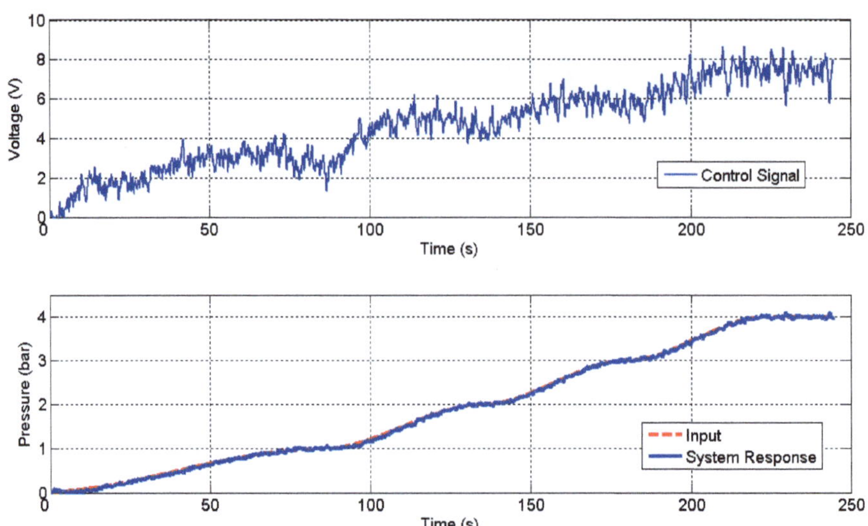

Fig. 4.33 Response of system to transition between trajectories

A reference trajectory as shown in Figure 4.33 was given to see success of the ACO-PID controller during the transition of system between trajectories with different slopes. System successfully carried out the transition between trajectorics with different slopes.

As is evident from the obtained graphics, in off-line study, real-time system control with the PID controller was optimized by ACO carried out successfully even if different references is applied.

4.4 Ziegler- Nichols (ZN) Method on PID Computation

In this method, the calculations are made with taking into account that system will make overshooting of 25% excess for step input. When a step input with certain value is applied to the system, this method is used if the system response is similar to the graph at Fig. 4.34.

PID coefficients are calculated according to the Table 4.2 with determining L and T values from the response of the system.

$$TF = \frac{Ke^{-Ls}}{Ts+1} \tag{4.8}$$

Here, L is delay time, T is time constant [63].

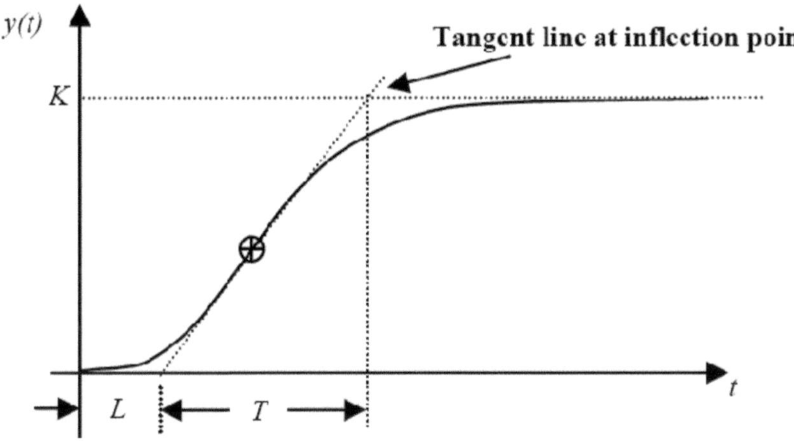

Fig. 4.34 Response of system to step input with constant amplitude

Table 4.2 Coefficient table of Ziegler-Nichols 1.method

Controller Type	Kp	Ki	Kd
P	$\dfrac{T}{L}$	∞	0
PI	$0.9\dfrac{T}{L}$	$\dfrac{L}{0.3}$	0
PID	$1.2\dfrac{T}{L}$	$2L$	$0.5L$

$$Ki = Kp\frac{1}{T_i} \tag{4.9}$$

$$Kd = KpT_d s \tag{4.10}$$

System response is shown at Figure 4.35, when a step input with constant amplitude is applied to the system. Here, when tangent to the turning point of the system at pressure filling curve is plot, time delay (L) and time constant (T) is identified as 5 s and 34 s respectively. PID constants are obtained as follows writing L and T values in place of Table 4.2.

$$Kp = 8.16$$

$$Ki = 0.1$$

$$Kd = 2.5$$

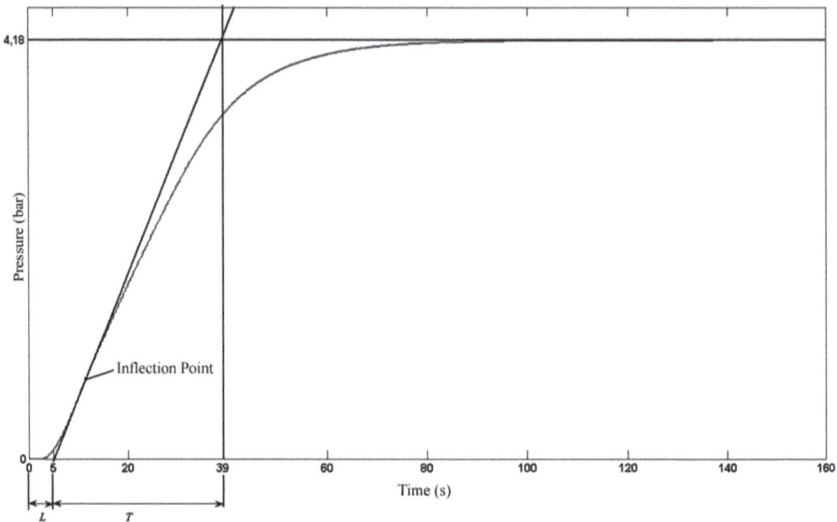

Fig. 4.35 Response of the system to step input

4.4.1   System Response to Step Inputs

Reference of 1 bar as shown in Figure 4.36 and 3 bar as shown in Figure 4.37 are given to system, then actual system outputs following these reference values are shown. Control signals providing these outputs are shown at the top of the figures. It can be understood from the behavior of the control signals, when system output approaches the reference value, PID controller was abandoned driving the system on maximum and was started to produce control signal for steady-state.

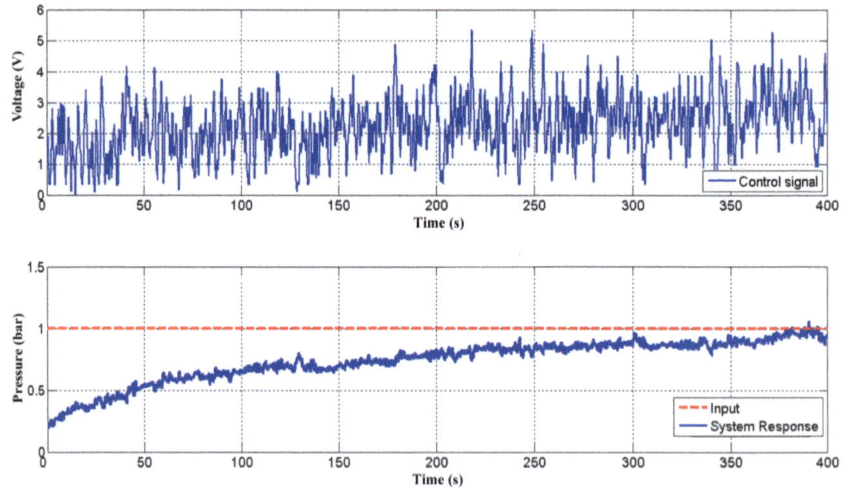

Fig. 4.36 Outputs of the Control and Actual System for reference input of 1 bar

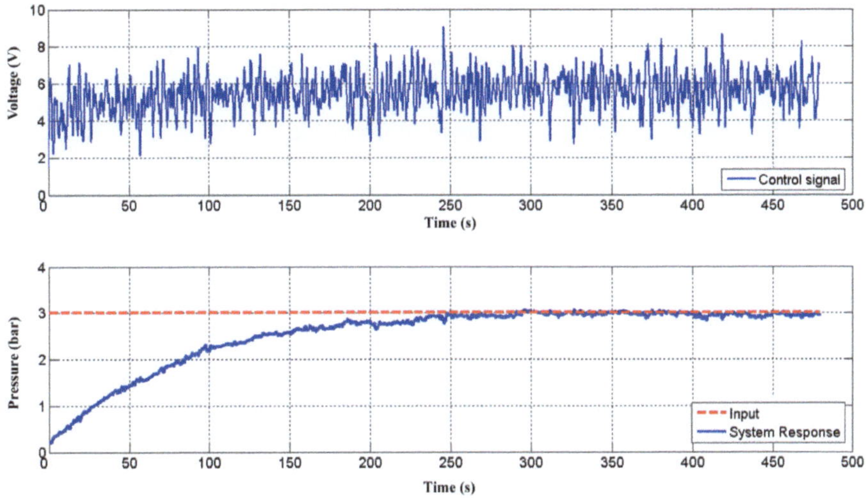

Fig. 4.37 Outputs of the Control and Actual System for reference input of 3 bar

4.4.2   System Behavior to the Transition between Step Inputs

In this section, to see the transition response of the ZN-PID controller to the different references, system response is investigated giving different references. After system with ZN-PID controller located the first reference value, transition of the system between the references are investigated. Graphics of this study are given at Figure 4.38. System can't carry out the transition between different inputs as shown in figure.

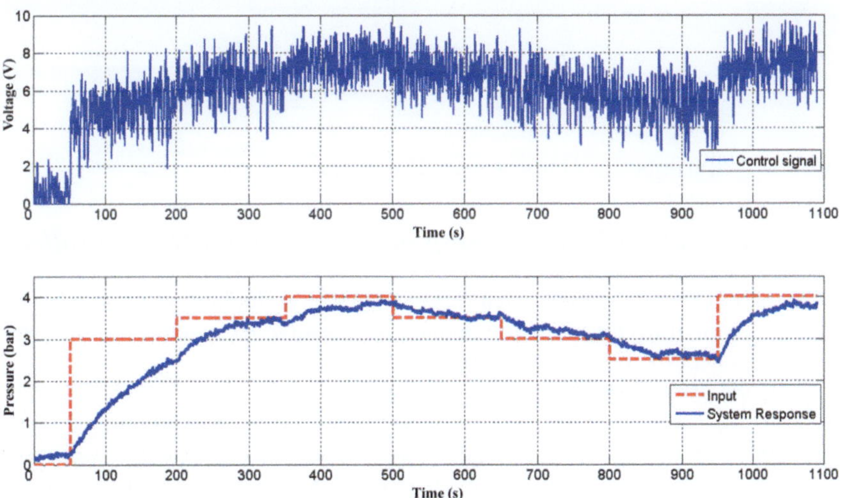

Fig. 4.38 Actual system response during step transitions

4.4.3 Behavior of the System When Distortion Was Imported to the System

In this part of the study performance of the purposed ZN-PID controller is carried out, in the case of a disturbance imported from outside.

During operation of the system, opening of the valve by 50% is provided by applying 6V voltage to pneumatic proportional pressure valve in the tank output. Thus, discharging of much air at the system is carried out. Disturbance has been imported by increasing the opening value of valve at the output. Discharge rate of the air in the tank increases by giving voltage more than 6 V. Thus, a negative disturbance has been imported to the system.

Behavior of the ZN-PID controller is shown at Figure 4.39 when 25% distortion is applied to the system. Disturbance input has applied for 120 s, but system has tracked the reference value with a steady state error. System output was again placed on reference value 180 s after disturbance is removed.

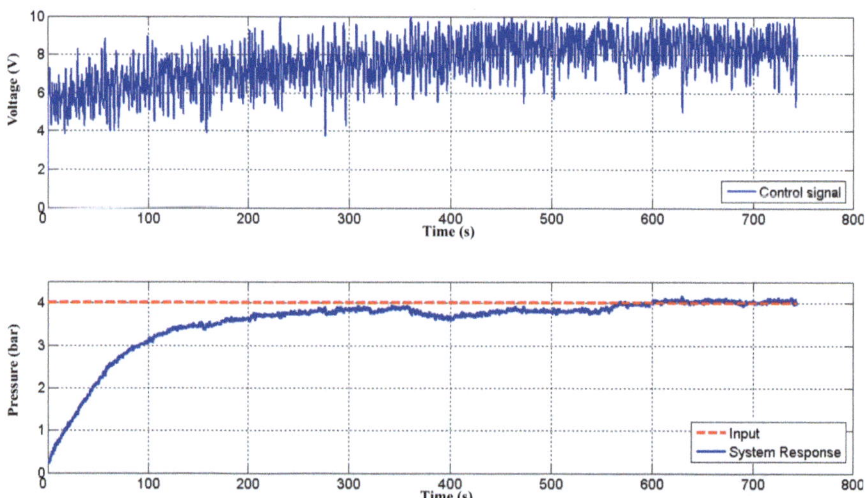

Fig. 4.39 System responses with applying %25 distortion

4.4.4 Trajectory Tracking Responses of the ZN-PID Controller

In this part of the study, the trajectories created with different grades from 0 bar to 4 bar (with the trajectory function defined in Equation 4.1 given to the system as reference trajectory the system behavior is investigated.

First, the system is purposed to follow from initial value to 4 bar. For this purpose, trajectory is generated with given 6 to the slope and given to the system as a reference. When system came to the 4 bar, response of the system was investigated during the discharge by creating the necessary trajectory for coming of system to the first state with same slope.

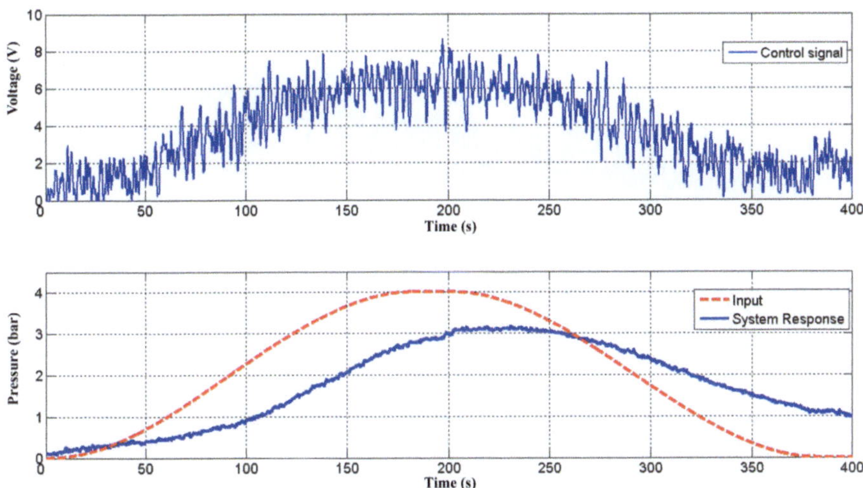

Fig. 4.40 Tracking of the reference trajectory with slope 6 by the system

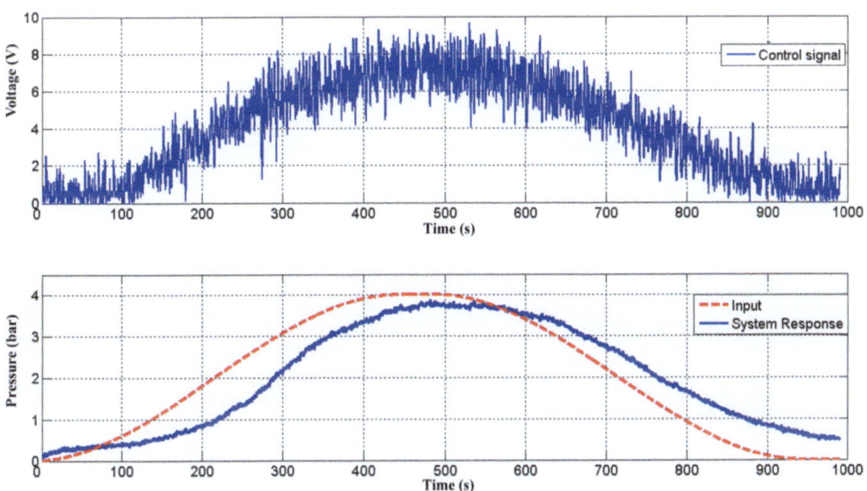

Fig. 4.41 Tracking of the reference trajectory with slope 15 by the system

Actual system behavior for tracking from the beginning to 4 bar with 6 and 15 slopes are shown at Figure 4.40 and 4.41, respectively. System with ZN-PID controller can't track the given trajectory as can be seen from the graphs.

A reference trajectory as shown in Figure 4.42 was given to see success of the ZN-PID controller during the transition of system between trajectories with different slopes. System couldn't carry out the transition between trajectories with different slopes.

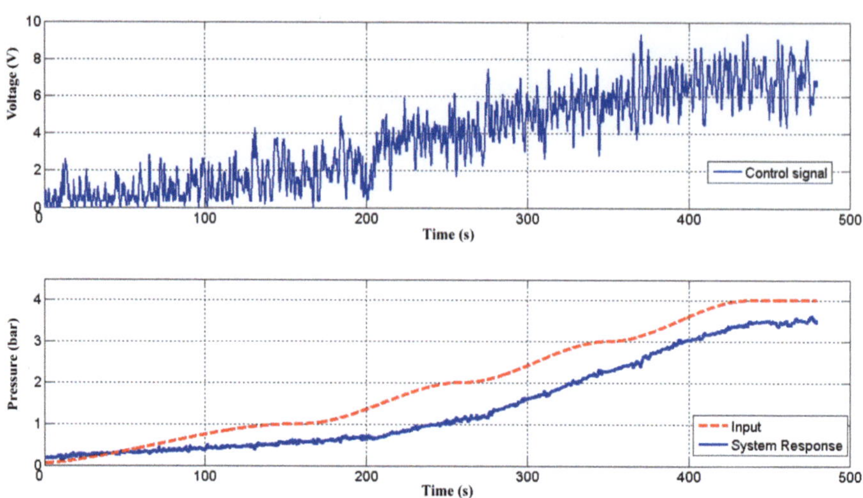

Fig. 4.42 System response to transition between trajectories

Chapter 5
Conclusion

In this paper, firstly dynamic model of pressure process system was obtained by NARX type ANN. Then validity of acquired model was proved by regression analysis $(R=0,997)$ and mean square error $(MSE=3,70.10^{-07})$. At this step, different ANN parameters were tested and ANN-30 model was found the good enough for modeling. Afterwards to prevent the fluctuations in the tank, a cubic trajectory function was used as an input reference signal. A cost function was designed to minimize the error along the trajectory for the GA-PID and ACO-PID controller. In order to compare results with traditional approach PID controller was also tuned Z-N methods. Outputs were evaluated by means of RMS error the respect to change of trajectory slope and reference pressure and robustness of controllers were also observed.

PID controller parameters were optimized by GA and ACO and calculated by ZN which is given in Table 3. The system responses with Genetic-PID, Ant-PID and ZN-PID controller are shown in Figure 5.1, when step inputs with different values according to PID coefficients shown Table 5.1 give to ANN model of the system.

Table 5.1 Obtained PID Controller Parameters with GA, ACO and ZN Methods

Controller	Kp	Ki	Kd
GA-PID	11,34	1,89	5,43
ACO-PID	12,88	1,44	3,92
ZN-PID	8,16	0,13	2,53

M. Ünal et al.: Optimization of PID Controllers Using AC and GA, SCI 449, pp. 69–72.
springerlink.com

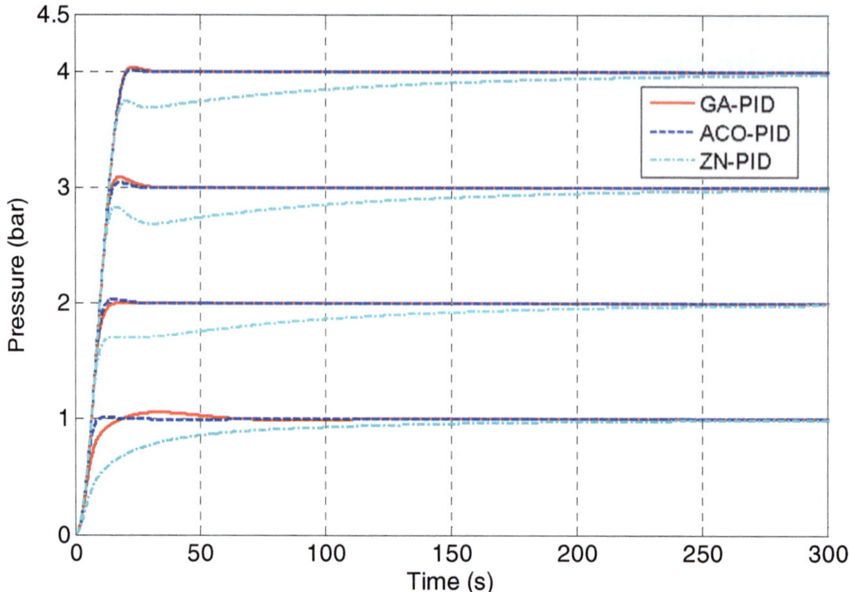

Fig. 5.1 System response to step inputs with different values

The performance of PID controllers obtained with GA, ACO and ZN is shown in Table 5.2 according to the amount of overshoot (OS), rise time (Tr), and settling time (Ts). Ant-ZN and Genetic-PID controller have given better results than ZN-PID.

Table 5.2 Control characteristics of PID controller

	Referans 1bar			Referans 2bar			Referans 3bar			Referans 4bar		
	GA	KKA	ZN	GA	KKA	ZN	GA	KKA	ZN	GA	KKA	ZN
Os (%)	5.7023	1.5162	-	0.1655	1.7152	-	3.0343	1.4994	-	0.9266	0.3261	-
Tr	9.338	5.0449	67.417	7.3431	6.8488	63.143	8.9744	9.0067	9.6084	12.475	12.475	12.682
Ts	60.973	16.438	253.54	12.781	11.143	200.65	18.236	13.734	144.27	18.066	18.067	95.128

Reference trajectories with 15 and 30° slopes have followed up with Genetic-PID, Ant-ZN-PID and PID controllers respectively. System responses are shown in Figure 5.2.

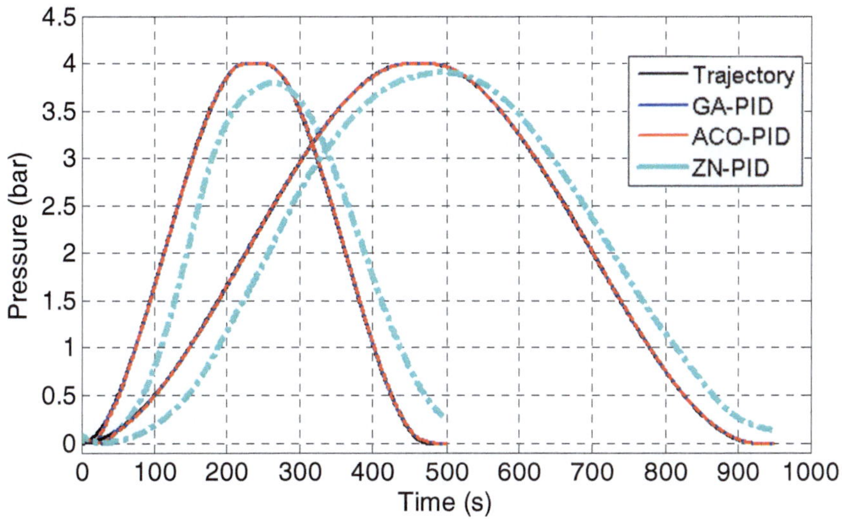

Fig. 5.2 System responses for different trajectories

RMS error between the system outputs - the reference trajectories shown Figure 5.2 are given in Table 5.3. Genetic-PID controller and Ant-PID controller have given more successful results then ZN-PID controllers.

Table 5.3 RMS errors for trajectory values

	RMS Errors		
Slope	**Genetik-PID**	**ANT-PID**	**ZN-PID**
15	0.0109	0.0143	0.4824
30	0.0056	0.0073	0.2755

With the evaluation of real-time operation of the system, it's seen that genetic-PID and Ant-PID controllers have successfully followed the references but ZN-PID controller has given the poor results compared to other controllers.

Thus, while the controller is designing for a real system due to obtaining the system model, it has shown that necessary control coefficients can be obtained during the experiments doing a large number of experiments and optimization techniques on a short duration.

While coefficients of the PID controller are obtaining with GA, the new values can be obtained from the previous values obtained with genetic operators in given search space. While optimization is doing with ACO, it produces the solution making the combination by searching the shortest path method in the given search space. Thus, the success PID coefficients for the system can be found. However, the new PID coefficients compared to other methods cannot be obtained in the optimization with ZN and there is not tending to the best solution. This feature is show the lack of the ZN comparison to the SC.

At the end of the numerous real time experiments, the following conclusions can be summarized:

1. NARX type ANN architecture can be used for the dynamic system modeling.
2. Reference pressure has no effect on the controller performance.
3. The RMS error decreased when trajectory slope was increased.
4. GA-PID and ACO-PID controllers show similar performance and their performances are better than ZN-PID controller at all reference pressure and trajectory values.
5. All designed controllers traced the defined trajectory within limited time
6. Finally, we concluded that both ACO and GA algorithms could be used to tune the PID controllers in the nonlinear system like pressure process easily with sufficient performance.

Appendix

A.1   Matlab Codes for ACO

A.1.1 Ant Colony Optimization Solver

```matlab
function [x,cval,time,assign]=ant(nvars,LB,UB,options)

% x          : independent variables for the cost function
% cval       : cost value
% time       : optimization time
% assign     : number of last tour
% nvars      : number of independent variables for the cost function
% LB         : lower bounds on the variables
(Lower specifies lower bounds as a vector)
% UB         : upper bounds on the variables
(Upper specifies lower bounds as a vector)
% options    : specify options for the Ant Colony Optimization Algorithm solver


%% Default options
defaultopt = struct('CostFcn',@ant_cost,...
'NumOfAnts', 20, ...
'Pheromone', 0.06, ...
'EvaporationParameter', 0.95, ...
'Positive Pheremone',0.2,...
'NegativePheremone',0.3,...
'MinValue', 0,...
'MaxValue',20,...
'Piece',101,...
'Figure','On',...
'MaxTour', 100);

%% Check input arguments and initialize them
% Check number of input arguments
errmsg = nargchk(1,4,nargin);
if ~isempty(errmsg)
    error([errmsg,' ANT requires at least 1 input argument.']);
end
if nargin< 4,  options = [];
if nargin< 3,  UB = ones(1,nvars);
if nargin< 2, LB = zeros(1,nvars);
end
```

```matlab
end
end

% Use default options if empty
if ~isempty(options) && ~isa(options,'struct')
   error('Options must be a valid structure.');
elseif isempty(options)
   options = defaultopt;
end
user_options = options;

% All inputs should be double
try
   dataType = superiorfloat(LB,UB);
if ~isequal('double', dataType)
   error('ANT only accepts inputs of data type double.')
end
catch
   error('ANT only accepts inputs of data type double.')
end
%% Start Ant Colony Optimization
clc        % clear command window
tic        % run chronometer
%% Initialize options
cost              = user_options.CostFcn;        % Cost function
ants              = user_options.NumOfAnts       % Number of ants
ph                = user_options.Pheromone       % Weight of pheromone
poz_ph            = user_options.PozitivePheromone
                   % Value of positivepheromone constant
neg_ph            = user_options.NegativePheromone
                   % Value of negative pheromone constant
lambda            = user_options.EvaporationParameter ;
                   % Evaporation Parameter
max_assign        =user_options.MaxTour          % Number of tour
piece             = user_options.Piece           % How many piece of design
variable matrices

%% Initialize variables
pher              = ones(piece,nvars);
indis             = zeros (ants,nvars);
costs             = zeros(ants,1);
cost_general      = zeros(max_assign,(nvars+1));

deger=zeros(piece,nvars);
deger(1,:)=LB;
for i=2:piece
```

```matlab
for j=1:nvars
    deger(i,j)=deger(i-1,j)+ (UB(j)-LB(j))/(piece-1);
end
end
%%
assign=0;
while (assign <max_assign)
for i=1:ants
% FINDING THE PARAMETERS OF VALUE
    prob = pher.*rand(piece,nvars);
for j=1:nvars
    indis(i,j) = find(prob(:,j) == max(prob(:,j)))   ;
end

    temp=zeros(1,nvars);
for j=1:nvars
    temp(j)=deger(indis(i,j),j);
end
    costs(i) = cost(temp);
% LOCAL UPDATING
    deltalocal = zeros(piece,nvars);
% CREATES THE MATRIX THAT WILL CONTAIN THE PHEREMONES
DEPOSITED FOR LOCAL UPDATING
for j=1:nvars
    deltalocal(indis(i,j),j)= (poz_ph *ph  / (costs(i)));
end
    pher = pher + deltalocal;
end
```

```matlab
% FIND THE BEST ANT AND UPDATE THE PHEROMONE OF THE BEST
ANT'S PARAMETERS
best_ant= min(find(costs==min(costs)));
worst_ant = min(find(costs==max(costs)));

deltapos = zeros(piece,nvars);
deltaneg = zeros(piece,nvars);

for j=1:nvars
    deltapos( indis(best_ant,j),j)  =  (ph /(costs(best_ant))) ;
        %UPDATING PHER OF nvars
    deltaneg( indis(worst_ant,j),j) = -(neg_ph * ph /(costs(worst_ant))) ;
        % NEGATIVE UPDATING PHER OF worst path
end

delta = deltapos + deltaneg;
pher = pher.^lambda + delta ;

assign=assign + 1;

% Update general cost matrix
for j=1:nvars
    cost_general (assign,j)=deger(indis(best_ant,j),j);
end
    cost_general (assign,nvars+1)=costs(best_ant);
%%
if strcmp(options.Figure,'On')
    set(gca,'xlim',[1,options.MaxTour],'Yscale','log');
    xlabelTour
    title('Change in Cost Value. Red: Means, Blue: Best')
    holdon
    plot(assign, mean(costs), '.r');
    plot(assign, costs(best_ant), '.b');
end
end
%%
list_cost =sortrows(cost_general,nvars+1);
for j=1:nvars
x(j)=list_cost(1,j);
end
    cval = ant_cost(x);
time = toc;
end
```

A.1.2 ACO Cost Function

```
function cval = ant_cost(x)
% Assign Kp, Ki and Kd
Kp = num2str(x(1));
Ki = num2str(x(2));
Kd = num2str(x(3));
% Run Close Loop Model with PID Controller
set_param('antz/PID','P', Kp ,'I', Ki, 'D', Kd);
sim('antz');
% Calculating the Cost Value
cval = mean(sqrt(power(reference - output,2)));
end
```

A.1.3 Run ACO

Simulink block diagram in Fig. A.1 is constituted to optimize the PID control with ACO. Process block represents the ANN model of the pressure process. Saturation object of 10V and a rate limiter to prevent sudden change of the proportional valve is placed to the PID controller output. Thus, the simulation is run with adjusting *Kp*, *Ki* and *Kd* values as the parameters of the controller while PID controller optimizing with ACO. The output and reference values obtained from program are sent to Workspace and used to calculate the cost value.

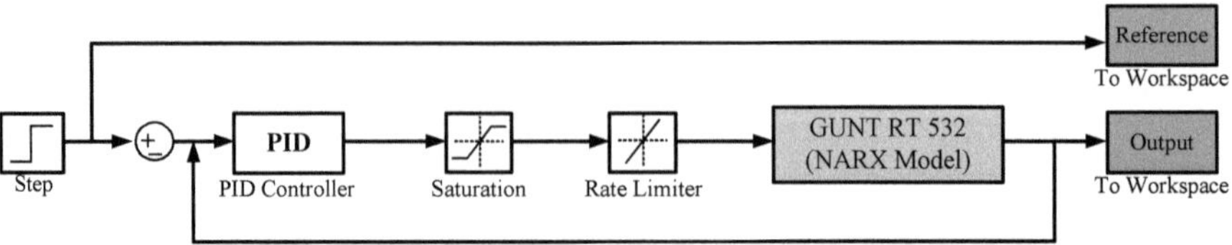

Fig. A.1 Close loop model with PID Controller for pressure process

```
clearall
clc
options = struct(  'CostFcn',@ant_cost,...
'NumOfAnts', 100, ...
'Pheromone', 0.06, ...
'EvaporationParameter', 0.95, ...
'PozitivePheremone',0.2,...
'NegativePheremone',0.3,...
'MaxTour', 100, ...
'MinValue', -10, ...
'MaxValue',10,...
'Figure','On',...
'Piece',501);
LB = [0  0 0];
UB = [20 1 10];
nvars=size(LB,2);
% Run ACO
[x,cval]=ant(nvars,LB,UB,options)
% Set PID Controller
Kp = num2str(x(1));
 Ki = num2str(x(2));
Kd = num2str(x(3));
% Run Close Loop Model with PID Controller
set_param('close_loop_model/PID','P', Kp ,'I', Ki, 'D', Kd);
sim('close_loop_model');
```

References

[1] Ünal, M.: Optimization of PID Controller Using Ant Colony / Genetic Algorithms and Control of The Gunt RT 532 Pressure Process, Master Master, Institue for Graduate Studies in Pure and Applied Sciences, Marmara University, İstanbul, Turkey (2008)

[2] Mok, H.S., Kim, G.T., Park, M.H., Rhew, H.W.: PI Controller Gains Tuning of The Pressure Control System By Open-Loop Frequency Response. IEEE (1988)

[3] Namba, R., Yamamoto, T., Kaneda, M.: Robust PID Controller and Its Application. IEEE (1997)

[4] Guo, C., Song, Q., Cai, W.: Real-time control of AHU based on a neural network assisted cascade control system. In: Robotics, Automation and Mechatronics, pp. 964–969 (2004)

[5] Jian, W., Wenjian, C.: Development of an Adaptive Neuro-Fuzzy Method For Supply Air Ressure Contro. In: Hvac System. IEEE (2000)

[6] Ünal, M., Erdal, H., Topuz, V., Ergüzel, T.T.: Isac Robot Kolunun Yörünge Takibi İçin PID Kontrolörün Genetik Algoritma ile Optimizasyonu. In: UMES, İzmit, Kocaeli (2007)

[7] Salami, M., Cain, G.: An Adaptive PID Controller Based on Genetic Algorithm Processor. In: Genetic Algorithms in Engineering Systems: Innovations and Applications (1995)

[8] Jones, A.H., DeMouraOliveira, P.B.: Genetic Auto-Tuning of PID Controllers. In: Genetic Algorithms in Engineering Systems: Innovations and Applications (1995)

[9] Gündoğdu, Ö.: Optimal-Tuning of PID Controller Gains Using Genetic Algorithms, Pamukkale Üniversitesi Mühendislik Fakültesi Mühendislik Bilimleri Dergisi, vol. 1 (2005)

[10] Dandıl, B., Gökbulut, M., Ata, F.: Genetik Algoritma ile Model Referans-PID Denetleyici Tasarımı. Fırat Üniversitesi Fen ve Mühendislik Bilimleri Dergisi 14, 33–43 (2002)

[11] Alli, H., Kaya, M.: Genetik Algoritma Kullanarak PID Kontrol Parametrelerinin Bulunması. Fırat Üniversitesi Fen ve Mühendislik Bilimleri Dergisi 13, 1–8 (2001)

[12] Mitsukura, Y., Yamamoto, T., Kaneda, M.: A Design of Self-Tuning PID Controllers Using a Genetic Algorithm. In: American Control Conference, pp. 1361–1365 (1999)

[13] Duan, H.-B., Wang, D.-B., Yu, X.-F.: Novel Approach to Nonlinear PID Parameter Optimization Using Ant Colony Optimization Algorithm. Journal of Bionic Engineering 3, 73–78 (2006)

[14] Ergüzel, T.T., Akbay, E.: ACO (Ant Colony Optimization) Algoritması ile Yörünge Takibi. In: UMES, İzmit, Kocaeli (2007)

[15] Varol, H.A., Bingul, Z.: A New PID Tuning Technique Using Ant Algorithm. In: American Control Conference, Boston, Massachusetts (2004)

[16] Fausett, L.: Fundamentals of neural networks. Prentice Hall Publishing, New York (1994)

[17] Zurada, J.M.: Introduction to Artificial Neural Systems. West Publishing Company, Minnesota (1992)

[18] Cavuto, D.J.: An Exploration and Development of Current Artificial Neural Network Theory and Applications with Emphasis of Artificial Life. Master Thesis, Citeseer (1997)

[19] Hagan, M.T., Demuth, H.B., Beale, M.: Neural Network Design. PWS, Boston (1997)

[20] Tsypkin, Y.Z., Nikolic, Z.J.: Foundations of the theory of learning systems. Academic Press, New York (1973)

[21] Onat, A., Kita, H., Nishikawa, Y.: Recurrent Neural Networks for Reinforcement Learning: Architecture, Learning Algorithms and Internal Representation. In: IEEE World Congress on Computational Intelligence (1998)

[22] Kecman, V.: Learning and Soft Computing: Support Vector Machines. In: Neural Networks, and Fuzzy Logic Models, pp. 121–191. MIT Press, Cambridge (2001)

[23] Ronco, E., Gawthrop, P.J.: Neural Networks for Modelling and Control. Centre for System and Control Department of Mechanical Engineering University of Glasgow (1997)

[24] Narendra, K.S., Parthasarathy, K.: Identification and Control of Dynamical Systems Using Neural Networks. IEEE Transactions on Neural Networks 1 (1990)

[25] Özçalık, H.R., Küçüktüfekçi, A.: Dinamik Sistemlerin Yapay Sinir Ağları ile Düz ve Ters Modellenmesi. KSÜ Fen ve Mühendislik Dergisi 6, 26–35 (2003)

[26] Şafak, K.K.: Identification of a Universal DC Motor Dynamics Using Feedforward Neural Networks. MS, Fen Bilimleri Enstitüsü, Boğaziçi Üniversitesi, İstanbul (1997)

[27] Coley, D.A.: An introduction to genetic algorithms for scientists and engineers. World Scientific Pub. Co. Inc. (1999)

[28] Quagliarella, D., Periaux, J., Poloni, C., Winter, G.: Genetic Algorithms and Evolution Strategy in Engineering Computer Science. John Wiley, NY (1998)

[29] Michalewicz, Z.: Genetic algorithms+ data structures = Evolution Programs. Springer, USA (1996)

[30] Goldberg, D.E.: Genetic algorithms in search, optimization, and machine learning. Addison-wesley (1989)

[31] Bäck, T.: Optimal Mutation Rates in Genetic Search. In: Proceeding of the Fifth International Conference on Genetic Algorithms, CL, USA (1993)

[32] Mitchell, M.: An introduction to genetic algorithms. The MIT Press (1998)

[33] Robertson, G.G.: Population Size in Classifier Systems. In: Proceeding of Fifth International Conference on Machine Learning, MA, USA (1988)

[34] Goldberg, D.E.: Sizing Populations for Serial and Parallel Genetic Algorithms. In: Proceeding of The Third International Conference on Genetic Algorithms, CL, USA (1989)

[35] Schaffer, D.J., Caruana, R., Eshelman, L., Das, R.: A Study of Control Parameters Affecting Online Perfpormance of Genetic Algorithms for Function Optimization. In: Proceeding of The Third International Conference on Genetic Algorithms, NJ, USA (1988)

[36] Janikow, C.Z., Michalewicz, Z.: An Experimental Comparison of Binary and Floating Point Representation in Genetic Algorithms. In: Proceeding of The Fourth International Conference on Genetic Algorithms, CL, USA (1991)

[37] Reeves, C.R.: Using Genetic Algorithms with Small Populations. In: Fifth International Conference on Genetic Algorithms, CL, USA (1993)

[38] Syswerda, G.: Uniform Crossover in Genetic Algorithms. In: Proceeding of the Third International Conference on Genetic Algorithms, NJ, USA (1989)

[39] Fogarty, T.C.: Varying the probability of Mutation in the Genetic Algorithm. In: Proceeding of The Third International Conference on Genetic Algorithms, NJ, USA (1989)

[40] Man, K.F., Tang, K.S., Kwong, S., Halang, W.A.: Genetic Algorithms for Control and Signal Processing (Advances in Industrial Control). Springer, England (1997)

[41] Goldberg, D.E., Deb, K., Korb, B.: Don't Worry be Messy. In: Proceeding of The Fourth International Conference on Genetic Algorithms. University of California, San Diego (1991)

[42] Kallel, L.S.: Alternative Random Initialization in Genetic Algorithm. In: Proceeding of The Seventh international Conference on Genetic Algorithms, MI, USA (1997)

[43] Buckles, B.P., Petry, E.F.: Genetic Algorithms. IEEE Computer Society Press, CL (1994)

[44] Michalewicz, Z.: Genetic algorithms + data structures = Evolution Programs. Springer, USA (1996)

[45] Grefenstette, J.J.: Optimization of Control Parameters for Genetic Algorithms. IEEE Transactions on Systems, Man and Cybernetics 16, 122–128 (1986)

[46] Goldberg, D.E., Sastry, K.: A practical schema theorem for genetic algorithm design and tuning. University of Illinois at Urbana IL, USA IlliGAL Report No 2001017 (2001)

[47] Yun, Y., Gen, M.: Performance Analysis of Adaptive Genetic Algorithms with Fuzzy Logic and Heuristics. Fuzzy Optimization and Decision Making 2, 161–175 (2003)

[48] Chambers, L.: Practical handbook of genetic algorithms, vol. 1-2. CRC Press, FL (1995)

[49] Dorigo, M.: Optimization, Learning and Natural Algorithms, PhD., Politecnico di Milano, Italie (1992)

[50] Colorni, A., Dorigo, M., Maniezzo, V.: Distributed Optimization by Ant Colonies. In: The First European Conference on Artificial Life, Paris, France (1992)

[51] Dorigo, M.: Ant Algorithms for Discrete Optimization. Artificial Life 5, 137 (1999)

[52] Hsiao, Y.-T., Chuang, C.-L., Chien, C.-C.: Ant Colony Optimization for Designing of PID Controllers. In: Intemational Symposium on Computer Aided Control Systems Design, Taiwan (2004)

[53] Bonabeau, E., Dorigo, M., Theraulaz, G.: Inspiration for Optimization From Social Insect Behaviour. Nature 406 (2000)

[54] Gambardella, L.M., Dorigo, M.: Ant-Q: A Reinforcement Learning approach to the traveling salesman problem. In: Twelfth Intern. Conf. on Machine Learning, ML 1995, pp. 252–260 (1995)

[55] Alaykıran, K., Engin, O.: Karınca Kolonileri Metasezgiseli ve Gezgin Satıcı Problemleri Üzerinde Bir Uygulaması. Gazi Üniv. Müh. Mim. Fak. Der. 20, 69–76 (2005)

[56] Can, B.: The optimal control of ITU Triga Mark II Reactor. Presented at the The Twelfth European Triga User's Conference, NRI Bucuresti-Pitesti, Romania (1992)

[57] Advantech PCI 1711 DAQ Card (December 30, 2009), `http://www.advantech.com/products/100-kS-s-12-bit-16-ch-SE-Input-PCI-Multifunction-Cards/mod_1-2MLGWA.aspx`

[58] Sanchez, E., Shibata, T., Zadeh, L.A.: Genetic Algorithms and Fuzzy Logic Systems: Soft Computing Perspectives (1997)

[59] Topuz, V.: Fuzzy Genetic Process Control. Ph.D, Institue for Graduate Studies in Pure and Applied Sciences, Marmara University, Istanbul, Turkey (2002)

[60] Yeniay, Ö.: An Overview of Genetic Algorithms. Anadolu Üniversitesi Bilim ve Teknoloji Dergisi 2, 37–49 (2001)

[61] Altiparmak, F., Dengiz, B., Smith, A.E.: An Evolutionary Approach For Reliability Optimization in Fixed Topology Computer Networks. Transactions on Operational Research 12, 57–75 (2000)

[62] Oliveira, P., Sequeira, J., Sentieiro, J.: Selection of Controller Parameters using Genetic Algorithms. Kluwer Academic Publishers, Dordrecht (1991)

[63] Ogata, K.: Modern Control Engineering, 3rd edn. Prentice Hall, New Jersey (1997)

Index

MIX
Papier aus verantwortungsvollen Quellen
Paper from responsible sources
FSC® C105338

If you have any concerns about our products,
you can contact us on
ProductSafety@springernature.com

In case Publisher is established outside the EU,
the EU authorized representative is:
**Springer Nature Customer Service Center GmbH
Europaplatz 3, 69115 Heidelberg, Germany**

Printed by Libri Plureos GmbH
in Hamburg, Germany